Springer
Tokyo
Berlin
Heidelberg
New York
Barcelona
Hong Kong
London
Milan
Paris
New Delhi
Singapore

K. Kuba, H. Higashida
D.A. Brown, T. Yoshioka (Eds.)

Slow Synaptic Responses and Modulation

With 112 Figures, Including 8 in Color

 Springer

Kenji Kuba, M.D., D.M.Sc.
Professor
Department of Physiology
Nagoya University School of Medicine
65 Tsurumai-cho, Showa-ku, Nagoya, Aichi 466-8850, Japan

Haruhiro Higashida, M.D., D.M.Sc.
Professor
Department of Biophysical Genetics
Kanazawa University, Graduate School of Medicine
13-1 Takara-machi, Kanazawa, Ishikawa 920-8640, Japan

David A. Brown, Ph.D.
F.R.S Professor
Department of Pharmacology
University College London, Gower Street, London, WC1E 6BT, England, UK

Tohru Yoshioka, Ph.D.
Professor
Department of Molecular Neurobiology
Waseda University, School of Human Science
Tokorozawa, Saitama 359-1192, Japan

QP
364
.5
.S56
2000

ISBN 4-431-70249-0 Springer-Verlag Tokyo Berlin Heidelberg New York

Library of Congress Cataloging-in-Publication Data
Slow synaptic responses and modulation /
   K. Kuba ... [et al.] eds.
      p. cm.
   Includes bibliographical references and index.
   ISBN 4431702490 (alk. paper)
   1. Neurotransmission. 2. Synapses. 3. Ion channels. I. Kuba, K. (Kenji), 1939–
QP364.5 .S56 1999
   573.8'5--dc21
                              99-38898

Typesetting: Camera-ready by editors
Printing and binding: Best-set Typesetter Ltd., Hong Kong
SPIN: 10681418

Preface

Information flow as nerve impulses in neuronal circuits is regulated at synapses. The synapse is therefore a key element for information processing in the brain. Much attention has been given to fast synaptic transmission, which predominantly regulates impulse-to-impulse transmission. Slow synaptic transmission and modulation, however, sometimes have been neglected in considering and attempting to understand brain function. Slow synaptic potentials and modulation occur with a considerable delay in response to the accumulation of synaptic and modulatory inputs. In these contexts, they are plastic in nature and play important roles in information processing in the brain.

A symposium titled "Slow Synaptic Responses and Modulation" was held as the satellite symposium to the 75th Annual Meeting of the Physiological Society of Japan on March 30–31, 1998, in Kanazawa. The theme was selected not only for the reason mentioned above, but also because of the considerable involvement of many Japanese scholars in establishing the basic issues.

Following the dawn of synaptic physiological research, as Sir John Eccles, Sir Bernard Katz, and Professor Stephen Kuffler carried out pioneer work, Professor Kyozou Koketsu and Professor Benjamin Libet, the students of Sir John Eccles, and their colleagues established the concept of slow synaptic responses and modulation by studying vertebrate sympathetic ganglia. Since then, the concept has been expanded with detailed investigations of both peripheral and central synapses at the levels of single ion channels, intracellular Ca^{2+} dynamics, intracellular transduction mechanisms, and genes.

This book explains first how the concept of slow synaptic responses and modulation emerged from studies on sympathetic ganglia, then how it was expanded into the elucidation of the detailed mechanisms in peripheral and central synapses, and finally how the recent progress in these issues was achieved. Major topics are: (1) postsynaptic mechanisms of slow synaptic potentials and modulation featuring a ubiquitous voltage-dependent K^+ channel, M-channel, and other K^+ channels, (2) presynaptic mechanisms of modulation focusing on voltage-dependent Ca^{2+} channels, Ca^{2+} release channels, and exocytotic machinery, (3) plasticity and development of synapses that are also regarded as modulation, (4) synaptic transmission in diseased or gene-targeted animals, and (5) many examples of modulation in a variety of synapses.

Each chapter begins with an explanatory review to provide the basic concepts and current understanding of the issues for students and those who are not familiar with the field, followed by recent findings presented by invited speakers and poster presenters in the symposium. The book is therefore relevant for students who are interested in synaptic physiology, and also for researchers and teachers working in the field.

We express our cordial thanks to the 75th Annual Meeting of the Physiological Society of Japan, the Japan Society for the Promotion of Science (for the "Signaling Mechanisms of Neurons" project in the "Research for the Future" program), the Ciba-Geigy Foundation (Japan), the Naito Foundation, the Nagoya University Foundation, the Saibikai Foundation of Kanazawa University Hospital, and the Yagi Foundation for their financial support to the Symposium and the publication of its results. We also thank the editorial staff of Springer-Verlag, Tokyo for their assistance in publication, Ms. Jyunko Hashimoto and Ms. Mayuko Mizutani for their help in preparation of the final version of all the manuscripts, and our colleagues for their help in organizing the symposium.

The Editors

Dr. Kyozo Koketsu, Emeritus Professor of Kurume University

Editors with Professor K. Koketsu after the symposium
(From left: K. Kuba, T. Yoshioka, K. Koketsu, D.A. Brown, and H. Higashida)

Contents

Ca^{2+} Currents and Modulation

Ca^{2+} Dynamics and Modulation

Exocytosis and Modulation

Synaptic Plasticity and Modulation

Synaptic Development, Modulation and Gene Expression

Synaptic Transmission and Modulation

Contributors

ABE, T. 247
AKAIKE, N. 46
AKAIKE, T. 446
AKANEYA, Y. 297
AKASU, T. 318, 375, 403
AKITA, T. 49, 173, 206, 271
ALBILLOS, A. 101
ALLEN, C.N. 200, 386
ASAI, T. 429
BITO, H. 182
BROWN, D.A. 15, 52
CARLEN, P.L. 67
CHOWDHURY, S.A. 442
CHUANG, H. 59
CHUHMA, N. 349
DE BARRY, J. 188
DEISSEROTH, K. 182
DOLPHIN, A.C. 121
DUPONT, J.L. 188
EGOROVA, A.B. 42
FUJITA, K. 320
FUJITA, R. 97
GANDÍA, L. 101, 130
GARCÍA, A.G. 101, 130
GOTO, M. 403
HAN, X.Y. 438
HASHII, M. 35, 209
HASHIMOTO, K. 268, 333
HASUO, H. 318, 403
HAUG, T. 78
HERNÁNDEZ-GUIJO, J.M. 130
HIGASHI, H. 397
HIGASHIDA, H. 35, 42, 46, 93, 161, 209, 367, 370
HOSHI, N. 42, 93, 161
HUANG, S.-M. 271

HUDA, K. 442
ICHIKAWA, K. 307
IKEDA, H. 429
IKEDA, M. 188, 200, 386
ILEVA, L. V. 307
INOKUCHI, H. 375, 397
IRIE, Y. 341
JAN, L.Y. 59
JAN, Y.N. 59
JANOSHAZI, A. 188
JIANG, Z.-G. 386
KAMIBAYASHI, K. 446
KAMIYA, H. 154
KANO, M. 268, 315, 325, 333
KASTIN, A.J. 161
KATO, H. 341
KAVALALI, E.T. 182
KAWAI, N. 416
KAWASAKI, S. 97
KAWASHIMA, T. 442
KIDOKORO, Y. 260
KIMURA, F. 297
KIMURA, S. 97
KINOSHITA, S. 297
KITAMURA, A. 271
KNIJNIK, R. 161
KOGURE, M. 266
KOJIMA, H. 307
KOKETSU, K. 1
KOSHIMOTO, H. 361
KUBA, K. 49, 158, 163, 173, 206, 212, 215, 271, 274, 320, 323
KUROMI, H. 260
KUZUYA, M. 158
LIPTON, S.A. 355
LIU, C. 323

The Dawn and Foundation of Slow Synaptic Potentials and Modulation

K. KOKETSU

Department of Physiology, Kurume University School of Medicine
67 Asahi-machi, Kurume 830-0011, Japan

Key words. Slow synaptic potential, Synaptic modulation, Sympathetic ganglia, Transmitter, Transmitter release, Receptor sensitivity, Resting potential, Action potential, Electrogenic Na^+-pump, Desensitization, ACh, Catecholamine, 5-HT, GABA, ATP.

Summary. The finding of slow potential changes, which were induced at the nonsynaptic membrane of ganglion cells and also preganglionic nerve terminals either by synaptic activation or by a direct application of various kinds of neurotransmitters in bullfrog sympathetic ganglia, led to the concept that the ganglionic transmission is modulated by the actions of many kinds of neurotransmitters released during ganglionic transmission.

Supporting this concept, the modulatory action of neurotransmitters on the ganglionic transmission, as well as other synaptic transmissions, had been demonstrated by a number of experiments reported from our laboratory since 1968. These results showed that a synaptic transmission is modulated by (1) changes in the amount of a transmitter released from presynaptic nerve terminals, (2) changes in the sensitivity of subsynaptic membrane to a transmitter, and (3) changes in the resting and/or action potential of postsynaptic as well as presynaptic membrane.

The outline of these experimental results reported between 1968 and 1988 has been reviewed here and the mode of modulatory actions of neurotransmitters and the mechanism underlying these modulatory actions discussed on the basis of electrophysiological standpoints.

I have been interested in the action of neurotransmitters modulating synaptic transmission for a long time, because these endogenous substances are responsible not only for the synaptic transmission itself but also for the modulation of synaptic transmission. The finding of slow synaptic potential changes, such as the slow excitatory postsynaptic potential (slow EPSP), the slow inhibitory postsynaptic potential (slow IPSP), and the late slow EPSP in bullfrog sympathetic ganglia

1

(Koketsu 1969), induced in the postganglionic cell by preganglionic nerve activation, led to the concept of synaptic modulation. Indeed, these slow potentials appeared to modulate the size of the fast EPSP representing ganglionic transmission by changing the resting potential level of postsynaptic membrane. In addition to these slow potential changes, we have been impressed by the many unique actions of various kinds of transmitters, which appeared to modulate the fast EPSP (Koketsu 1969), since we found that preganglionic nerve terminals were depolarized by the action of acetylcholine (ACh) (Koketsu and Nishi 1968), and we expected that such a depolarization of presynaptic nerve terminals would modulate the amount of ACh released from presynaptic membrane. I present here the outline of our research work, which has been carried out in our laboratory between 1968 and 1988 and demonstrated the different modes of modulatory actions of many kinds of transmitters on synaptic transmissions (Koketsu 1969, 1981, 1984, 1987; Koketsu and Akasu 1986; Koketsu et al. 1991; Kuba and Koketsu 1978).

The efficiency of sympathetic ganglionic transmission is represented by the size of the fast EPSP. In general, the efficiency of synaptic transmission is represented by the size of the postsynaptic potential. The size of the postsynaptic potential is determined by (1) the amount of a transmitter released from the presynaptic membrane, (2) the sensitivity of subsynaptic membrane to a particular transmitter, and (3) the resting membrane potential and/or conductance of postsynaptic membrane.

The modulation of a synaptic transmission induced by a particular transmitter would be observed by changes in the size of the postsynaptic potential, which are caused by the action of transmitters other than this particular transmitter. Thus, such a modulatory action of transmitters would be expected to be achieved by three different modes, namely (1) modulatory actions on the amount of a released transmitter, (2) modulatory actions on the sensitivity of the subsynaptic membrane to released transmitter, and (3) modulatory action on the resting membrane potential and/or conductance of postsynaptic membrane. On the basis of such a expectation, the modulatory action of neurotransmitters on these three different factors determining the size of the fast EPSP was extensively investigated in our laboratory.

In connection with the expected mode 3, it had been considered that the membrane of a neurone is composed of (1) the chemically excitable membrane and (2) the electrically excitable membrane. Furthermore, it had been considered for long time that a transmitter is supposed to act only on the chemically excitable membrane but not on the electrically excitable membrane. In other words, transmitters had been considered to be able to act only on the subsynaptic membrane at synaptic junctions. This concept, however, was reconsidered in the experiments where the slow postsynaptic potentials, such as the slow EPSP, slow IPSP, and late slow EPSP, were found in the bullfrog's sympathetic ganglion cell membrane, where no synaptic junctions could be assumed to exist (Koketsu 1969). Furthermore, the experimental finding that slow depolarization is produced by the action of ACh at the bullfrog's preganglionic nerve fiber terminals, where no cholinergic synaptic junctions apparently exist, strongly supported this consideration (Koketsu and Nishi 1968). Indeed, many experimental results, which had been accumulated during about 15 years since 1968 in our laboratory, clearly suggested that many transmitters act not only to the chemically excitable membrane but also to the electrically excitable membrane producing action potentials. This led to a suggestion that there are receptors distributed on the nonsynaptic membrane, as well as on the subsynaptic membrane in a neurone. Accordingly, this suggestion supported the concept that transmitters are capable of modulating not only the resting potential but also the action potential.

It would be thus finally concluded, on the basis of the concept described above, that the modulatory action of transmitters on a synaptic transmission can be divided into the following the three modes: (1) modulatory actions on the amount of a released transmitter; (2) modulatory actions on the sensitivity of the subsynaptic membrane (chemically excitable membrane) to a released transmitter; and (3) modulatory actions on the resting and/or action potential of the postsynaptic as well as presynaptic membrane (electrically excitable membrane). Experimental results which demonstrated these modulatory actions of transmitters and were reported from our laboratory are briefly reviewed here.

1 Modulatory Actions on the Amount of a Released Transmitter

The first recording of the membrane depolarization of cholinergic nerve terminals in vertebrates was made from bullfrog sympathetic preganglionic nerve terminals as a response to the nicotinic action of ACh (Koketsu and Nishi 1968). It was thus suggested that the release of ACh from preganglionic nerve terminals was modulated by the nicotinic action of ACh (Nishi 1970). The depolarization induced by γ-aminobutyric acid (GABA) was also demonstrated in the same preparation by Koketsu et al. (1974). Subsequently, Kato et al. (1978) have demonstrated that the release of ACh is actually depressed in the presence of GABA. These experimental observations facilitated further investigations for examining the hypothesis that the release of ACh is modulated under the effects of neurotransmitters, including ACh itself and other neurotransmitters. Indeed, a considerable body of evidence has accrued consistent with ACh release being actually modulated by many kinds of neurotransmitters (Koketsu and Akasu 1985; Koketsu et al. 1982a).

ACh is released from the presynaptic membrane when the action potential arrives at cholinergic nerve terminals. Such an ACh release triggered by action potentials is called active ACh release, as compared with a spontaneous release of ACh, which is a random release of ACh quanta from presynaptic nerve terminals at a resting state. The mechanism underlying the ACh release is not fully understood at present. Nevertheless, the quantum theory, which was proposed by

Katz and his collaborators (Katz 1966), has provided a powerful research basis for investigating the mechanism of ACh release.

Although presynaptic nerve terminals of sympathetic ganglia were found to be depolarized by the nicotinic action of ACh, the implication of such a action of ACh was difficult to solve because no conclusive experimental evidence had appeared to suggest that the nicotinic action of ACh indeed changes either spontaneous or active release of ACh from these nerve terminals. It was found, however, that the muscarinic action of ACh actually causes a reduction of the quantal content of the fast EPSP in sympathetic ganglia (Koketsu and Yamada 1982).

Kato et al. (1978) demonstrated that GABA significantly decreased the quantal content of the fast EPSP. In the superior cervical ganglion of rat, GABA depolarized the preganglionic nerve fibers and reduced the amount of ACh release from preganglionic nerve terminals (Brown and Higgins 1979). It was suggested, on the other hand, that the reduction of ACh release by GABA is due to an increase of the membrane conductance of the presynaptic nerve terminals by the shunting effect of chloride conductance changes (Kato and Kuba 1980).

Adrenaline has been known to facilitate the ACh release from cholinergic nerve terminals in the neuromuscular junction (Kuba 1970; Kuba and Tomita 1971). On the other hand, in superior cervical ganglia, adrenaline or dopamine decreased both the frequency of the miniature EPSP and the quantal content of the fast EPSP (Christ and Nishi 1971; Dun and Nishi 1974). Tsurusaki (1987) found that adrenaline depresses the amount of ACh released from presynaptic nerve terminals in vesical parasympathetic ganglia. In bullfrog sympathetic ganglia, it appears that adrenaline in small concentrations increases ACh release, while in relatively high concentrations reduces it (Koketsu 1981; Kuba and Kumamoto 1986). It was thus suggested that adrenaline has bimodal modulatory actions, namely, facilitatory and inhibitory regulation of ACh release at the nerve terminals (Koketsu 1981). It was further suggested that the inhibitory action of adrenaline on the ACh release is not due to the change in the action potential of preganglionic nerve terminals (Kato et al. 1985).

The presynaptic effects of 5-hydroxytryptamine (5-HT) are comparable to those observed with adrenaline (Hirai and Koketsu 1980). The facilitatory and the inhibitory actions of 5-HT on ACh release from preganglionic nerve terminals are concentration dependent. Namely, 5-HT in relatively small concentrations increases the amount of ACh release, whereas in relatively high concentrations it decreases the amount of ACh release from preganglionic nerve terminals in bullfrog sympathetic ganglia. Bimodal actions of 5-HT were also observed in the frog neuromuscular junction (Hirai and Koketsu 1980). In parasympathetic ganglia, 5-HT depresses ACh release, acting on the presynaptic 5-HT subtype receptor (Nishimura et al. 1988a), and the depression of ACh release is followed by a long-lasting facilitation of ACh release (Nishimura et al. 1988b). Histamine was also found to show a similar concentration-dependent dual action on the ACh release from preganglionic nerve terminals (Yamada et al. 1982).

Adenosine triphosphate (ATP) was proposed to decrease the amount of ACh release from preganglionic nerve terminals in bullfrog sympathetic ganglia (Nakamura et al. 1974). This was confirmed by the subsequent experiment demonstrating that the quantal content of the fast EPSP was markedly depressed in the presence of ATP (Akasu et al. 1982, 1983a,b). It was also found that the membrane of preganglionic nerve terminals was markedly depepolarized by the action of ATP (Akasu et al. 1982). Thus, it was suggested that the ATP-induced reduction of the ACh release was presumably caused by the depolarization of the preganglionic nerve terminals.

Substance P, an endogenous neuropeptides, was found to increase the ACh release from motor nerve terminals (Akasu 1986b; Kojima-Nishimura and Akasu 1983). Substance P showed a dual action on the release of ACh from bullfrog sympathetic preganglionic nerve terminals, namely, it facilitated the ACh release in low concentrations, while it depressed the ACh release in high concentrations (Akasu et al. 1983f).

Luteinizing hormone-releasing hormone (LHRH) in small concentrations was found to facilitate the release of ACh from motor nerve endings (Akasu 1986a; Kojima et al. 1982). The quantal release of ACh from preganglionic nerve terminals was reduced by LHRH in bullfrog sympathetic ganglia (Hasuo and Akasu 1986). It is interesting that the quantal content of fast EPSP was strongly depressed (Hasuo and Akasu 1988a,b) during the production of the late slow EPSP, which presumably was induced by the action of LHRH-like peptide (Jan and Jan 1982; Katayama and Nishi 1982). A brief bath-application of cholecystokinin octapeptide (CCK-8) produced a depression, followed by a long-lasting facilitation, of the amplitude and the quantal content of the end-plate potential in the frog neuromuscular junction (Akasu et al. 1986). Kyotorphin increased the ACh release from preganglionic nerve terminals in bullfrog sympathetic ganglia (Hirai and Katayama 1985).

Two interesting phenomena relating to the modulation of release of neurotransmitter, which were found in our laboratory, are described next. One interesting phenomenon was a long-term potentiation induced by the action of adrenaline in bullfrog sympathetic ganglia. This is an augmentation of ACh release following a brief treatment of the ganglion with adrenaline in concentrations which depress the ACh release (Kuba et al. 1981). Upon removal of adrenaline from perfusate, both the amplitude and quantal content of the fast EPSP increase progressively until they reach a maximum value within approximately 40 min, with the facilitation lasting for 2-5 h (Kuba and Kumamoto 1986; Kuba et al. 1981; Kumamoto and Kuba 1983, 1987). Another interesting phenomenon was antidromic inhibition of ACh release; namely, the release of ACh is inhibited by antidromic activation of sympathetic ganglion cells (Miyagawa et al. 1981). Most likely, adrenaline was released from ganglion cells activated antidromically (Suetake et al. 1981) and exerted its action on the preganglionic nerve terminals.

The mechanism underlying the modulation of the release of a transmitter by other transmitters was not entirely clarified. On the basis of electrophysiology, the amount of released transmitter would be determined by the resting potential and/or conductance and by the action potential of presynaptic membrane. It may be thus suggested that the modulation of the amount of a released transmitter, at least a part of it, may be caused by the changes in resting or action potentials in presynaptic nerve terminals.

2 Modulatory Actions on the Sensitivity of Subsynaptic Membrane to a Released Transmitter

The sensitivity of membrane receptors can be estimated by recording membrane responses induced by the action of a constant amount of a particular neurotransmitter. If the sensitivity is changed in the presence of some kind of another transmitter, the amplitude of recorded membrane responses would be changed, demonstrating that the sensitivity of receptors to this particular neurotransmitter is modulated by the action of another transmitter.

In 1968, the amplitude of the ACh-potential induced by iontophoretic application of ACh to the endplate was reported to be decreased in the presence of 5-HT (Colomo et al. 1968). This finding indeed demonstrated that the sensitivity of the endplate to ACh could be modulated by the action of 5-HT. In 1981, this observation was confirmed in our laboratory, reporting that 5-HT interacts directly with nicotinic ACh-receptors and thereby depresses the sensitivity of these receptors to ACh, in sympathetic ganglion cells as well as muscle endplates (Akasu et al. 1981a). Such a depressant action of 5-HT on the sensitivity of the ACh-receptor was suggested to be due to decrease of the affinity of ACh for ACh-receptor sites, in a manner similar to d-TC, which blocks nicotinic transmission in a competitive manner (Akasu and Koketsu 1986; Akasu et al. 1981a, 1983c; Koketsu et al. 1982a).

In 1982, Koketsu et al. (1982b) proposed the concept that the sensitivity of ACh-receptors is modulated by the action of catecholamines. In this experiment, catecholamine was suggested to depress the sensitivity of ACh-receptors in a noncompetitive manner. The difference of the mode of modulatory action on ACh-receptor sensitivity, being observed between the action of 5-HT and catecholamine, was confirmed by the following experiments (Koketsu et al. 1982a).

In 1973, Scuka (1973) observed that histamine reduced the amplitude of ACh-potential at frog skeletal muscle. This observation was evaluated in the experiments carried out in our laboratory, in which histamine depresses the ACh-current in a competitive manner (Ariyoshi et al. 1984, 1985; Ohta et al. 1984). LHRH was found to depress the sensitivity of nicotinic receptors in a noncompetitive manner, similar to catecholamine (Akasu et al. 1983e). Akasu et al. (1983d) reported that P-substance depressed the sensitivity of ACh-receptors in a manner similar to that of LHRH.

The sensitivity of the ACh-receptor is modulated, not only by the depressant action of neurotransmitters (endogenous antagonists), but also by facilitatory action of neurotransmitters (endogenous sensitizer). This concept was supported by the experimental evidence that the sensitivity of ACh-receptor is augmented by the action of ATP. The possibility that the sensitivity of nicotinic ACh-receptors can be increased by the action of ATP was first suggested by Ewald (1976). This possibility was confirmed in our laboratory, and it was suggested that ATP increased the receptor sensitivity without changing the affinity of ACh for its recognition site (Akasu and Koketsu 1985; Akasu et al. 1981b). Potentiation of receptor sensitivity by transmitters has been observed also at other receptors. ATP reversibly augmented the depolarization induced by iontophoretic application of GABA onto bullfrog primary afferent neurones (Morita et al. 1984).

3 Modulatory Actions on the Resting and Action Potential of Electrically Excitable Membrane

3.1 Changes in the Resting Potential by a Transmitter

3.1.1 Changes in the Passive Ion Transport

The slow EPSP, slow IPSP, and late slow EPSP may be produced by changes in the passive ion transport in sympathetic ganglion cells. These potential changes may be initiated from the electrically excitable membrane, where no synaptic connections (viz., no subsynaptic membrane) exist, and may be responsible for modulation of the fast EPSP.

Depolarization of preganglionic nerve terminals induced by direct action of ACh was demonstrated in 1968 (Koketsu and Nishi 1968). Such a depolarization would reduce the amount of released ACh (presynaptic modulation). Similar ACh depolarization, which is different from the fast EPSP, was suggested to exist in postganglionic neurone (Koketsu 1969).

It has been well known that many kinds of transmitters other than ACh are capable to change the resting membranes potential in both pre- and postganglionic cell membranes where no synaptic connections exist (Kuba and Koketsu 1978).

3.1.2 Changes in the Active Ion Transport

The existence of a cellular regulatory mechanism, by which neuronal substances exhibit a physiological role through modulation of Na^+-K^+ pump activity, was a general concept (Koketsu 1971). The fact that many kinds of transmitters are able to change the resting membrane potential by affecting the active ion transport, such as the electrogenic Na^+-pump, has been suggested in a number of experiments reported from our laboratory.

The amplitude of membrane potential changes induced by an activation of the electrogenic Na^+-pump would be estimated by recording the K^+-activated hyperpolarization elicited by a direct quick application of K^+ ions to the membrane of preparations that   were perfused constantly by K^+-free solution. The K^+-activated hyperpolarization was markedly facilitated in the presence of adrenaline (Akasu and Koketsu 1976a,b) and ACh (Minota and Koketsu 1979).

The synaptically induced K^+-activated hyperpolarization (Kuba and Koketsu 1978) was found to be enhanced in the presence of adrenaline as well as 5-HT (Shirasawa et al. 1976). In skeletal muscle, it had been suggested that catecholamines stimulate the Na^+-pump and that this may cause the membrane hyperpolarization of these muscles (Kuba and Koketsu 1978). It was demonstrated in our laboratory that these muscle cells produced K^+-activated hyperpolarization and that adrenaline clearly augmented this hyperpolarization (Koketsu and Ohta 1976). It was also found that K^+-activated hyperpolarizations of bullfrog atrium muscle fibers (Akasu et al. 1977) and bullfrog visceral nerve fibers (Morita and Koketsu 1979) were enhanced under the effect of adrenaline. Furthermore, the experimental evidence showing that the electrogenic Na^+-pump current actually augmented during the K^+-activated hyperpolarization of heart muscle was seen in the presence of adrenaline (Akasu et al. 1978).

It appeared that adrenaline increases neither the total number of pumping sites nor the Na^+/K^+-coupling ratio, and that adrenaline increases the rate of Na^+ extrusion by increasing the overall affinity of the pumping sites to extracellular K^+ in a fixed concentration (Akasu et al. 1978; Morita and Koketsu 1979). It was further suggested by a subsequent experiment that β-adrenoceptors have an important role in modulating Na^+-K^+ pump activity and that the Na^+-K^+ ATPase and β-adrenoceptors may be the functional unit which operates the membrane machinery driving the Na^+-K^+ pump (Kaibara et al. 1985).

3.2 Changes in the Action Potential by a Transmitter

The modulatory action of transmitters on the action potential of a neurone was first observed in 1974 in the experiment in which the action potential, particularly its after hyperpolarization, of sympathetic ganglion cells was observed to be depressed by the muscarinic action of ACh (Koketsu 1974,1984). It was suggested in this experiment that the action potential of neurones, and consequently the voltage-dependent ion channels on the neurone membrane, could be physiologically modulated by neurotransmitters, similar to cardiac muscle fibers. This observation was reexamined in detail and confirmed by the following experiments (Kuba and Koketsu 1975, 1976; Morita et al. 1982). The calcium-dependent potassium conductance was also found to be depressed by the action of muscarinic agonists (North and Tokimasa 1983; Tokimasa 1984).

Subsequent experiments demonstrated that the Ca^{2+}-potential of sympathetic ganglion cells was also modulated by the muscarinic action of ACh (Koketsu 1984). These experiments demonstrated that the voltage-dependent K^+-current as well as the voltage-dependent Ca^{2+}-current of sympathetic ganglion cells can be modulated by the muscarinic action of ACh.

The fact that the action potential of sympathetic ganglion cells is modulated by the action of catecholamines was also observed in the experiments carried out in our laboratory (Koketsu and Minota 1975). It was further demonstrated that the voltage-dependent K^+ and Ca^{2+} currents and also the voltage-dependent Na^+-current were modulated by the action of adrenaline (Minota and Koketsu 1977).

Voltage-clamp experiments carried out with sympathetic ganglion cells demonstrated that the so-called M-current is depressed, not only by the muscarinic action of ACh (Adams et al. 1982; Brown and Adams 1980), but also by polypeptides, such as LHRH (Adams et al. 1982; Akasu et al 1983h) and substance P (Akasu et al. 1983g). The M-current was also markedly depressed by ATP (Akasu et al. 1983a). The delayed rectifier K^+-current was reported to be depressed by ACh (Akasu and Koketsu 1981, 1982), catecholamines (Akasu and Koketsu 1981; Koketsu and Akasu 1982), LHRH (Akasu et al. 1983h) and substance P (Akasu et al. 1983g).

The depression of the slow inward Ca^{2+}-current was caused by catecholamine (Akasu and Koketsu 1981; Koketsu and Akasu 1982) and ACh (Akasu and Koketsu 1981, 1982). Voltage-clamp experiments also suggested that the fast inward Na^+-current is modulated by ACh (Akasu and Koketsu 1982), catecholamine (Koketsu and Akasu 1982), LHRH (Akasu et al. 1983h) and substance P (Akasu et al. 1983g).

An interesting property of receptors is desensitization to a specific agonist. Interestingly, the action of neurotransmitters on the voltage-dependent ionic channels seems to display desensitization, which would be never observed under comparable conditions with pharmacological or toxicological agents. Indeed, action potentials of cardiac muscle fibers, being depressed by ACh, were found to display desensitization when the application of ACh was sustained (Tokimasa et al. 1980, 1981). The slow inward Ca^{2+}-current recorded by voltage-clamp experiments from atrial muscle fibers also displayed desensitization during a sustained muscarinic action of ACh (Hasuo et al. 1982). Similar experiments on bullfrog sympathetic ganglion cells have led to the conclusion that the muscarinic receptors controlling the voltage-dependent K^+ and Ca^{2+}-currents display desensitization during a prolonged application of ACh (Akasu and Koketsu 1980). It was also found that neuropeptides facilitate the desensitization of nicotinic

ACh-receptor in frog skeletal muscle end-plate (Akasu et al. 1984). These results suggest that the voltage-dependent ion channel is composed of a macromolecular complex with a receptor with which the transmitter interacts.

References

Adams PR, Brown DA, Constanti A (1982) Pharmacological inhibition of the M-current. J Physiol (Lond) 332:223-262

Akasu T (1986a) Luteinizing hormone releasing hormone modulates the cholinergic transmission in frog neuromuscular junction. Jpn J Physiol 36:25-42

Akasu T (1986b) The effects of substance P on neuromuscular transmission in the frog. Neurosci Res 3:275-284

Akasu T, Koketsu K (1976a) Adrenaline and the electrogenic sodium pump in *Rana catesbeiana* sympathetic ganglion cells. Experientia 32:57-59

Akasu T, Koketsu K (1976b) The effect of adrenaline on the K^+-activated hyperpolarization of the sympathetic ganglion cell membrane in bullfrogs. Jpn J Physiol 26:289-301

Akasu T, Koketsu K (1980) Desensitization of the muscarinic receptor controlling action potentials of sympathetic ganglion cells in bullfrogs. Life Sci 27:2261-2267

Akasu T, Koketsu K (1981) Modulatory actions of neurotransmitters on voltage-dependent membrane currents in bullfrog sympathetic neurones. Kurume Med J 28:345-348

Akasu T, Koketsu K (1982) Modulation of voltage-dependent currents by muscarinic receptor in sympathetic neurones of bullfrog. Neurosci Lett 29:41-45

Akasu T, Koketsu K (1985) Effect of adenosine triphosphate on the sensitivity of the nicotinic acetylcholine-receptor in the bullfrog sympathetic ganglion cell. Br J Pharmacol 84:525-531

Akasu T, Koketsu K (1986) 5-Hydroxytryptamine decreases the sensitivity of nicotinic acetylcholine receptor in bull-frog sympathetic ganglion cells. J Physiol (Lond) 380:93-109

Akasu T, Ohta Y, Koketsu K (1977) Activation of electrogenic Na^+ pump by epinephrine in bullfrog atrium. Jpn Heart J 18:860-866

Akasu T, Ohta Y, Koketsu K (1978) The effect of adrenaline on the electrogenic Na^+ pump in cardiac muscle cells. Experientia 34:488-490

Akasu T, Hirai K, Koketsu K (1981a) 5-Hydroxytryptamine controls ACh-receptor sensitivity of bullfrog sympathetic ganglion cells. Brain Res 211:217-220

Akasu T, Hirai K, Koketsu K (1981b) Increase of acetylcholine-receptor sensitivity by adenosine triphosphate: a novel action of ATP on ACh-sensitivity. Br J Pharmacol 74:505-507

Akasu T, Hirai K, Koketsu K (1982) Modulatory effect of ATP on the release of acetylcholine from presynaptic nerve terminals in bullfrog sympathetic ganglia. Kurume Med J 29:75-83

Akasu T, Hirai K, Koketsu K (1983a) Modulatory actions of ATP on membrane potentials of bullfrog sympathetic ganglion cells. Brain Res 258:313-317

Akasu T, Hirai K, Koketsu K (1983b) Modulatory actions of ATP on nicotinic transmission in bullfrog sympathetic ganglia. In: Daly JW, Kuroda Y, Phillis JW, et al (eds) Physiology and pharmacology of adenosine d erivatives. Raven Press, New York, pp 165-171

Akasu T, Karczmar AG, Koketsu K (1983c) Effects of serotonin (5-hydroxytryptamine) on amphibian neuromuscular junction. Eur J Pharmacol 88:63-70

Akasu T, Kojima M, Koketsu K (1983d) Substance P modulates the sensitivity of the nicotinic receptor in amphibian cholinergic transmission. Br J Pharmacol 80:123-131

Akasu T, Kojima M, Koketsu K (1983e) Luteinizing hormone-releasing hormone modulates nicotinic ACh-receptor sensitivity in amphibian cholinergic transmission. Brain Res 279:347-351

Akasu T, Kojima M, Koketsu K (1983f) Modulatory effect of luteinizing hormone-releasing hormone and substance P on the nicotinic transmission in bullfrog sympathetic ganglia. J Physiol Soc Jpn 45:418

Akasu T, Nishimura T, Koketsu K (1983g) Substance P inhibits the action potentials in bullfrog sympathetic ganglion cells. Neurosci Lett 41:161-166

Akasu T, Nishimura T, Koketsu K (1983h) Modulation of action potential during the late slow excitatory postsynaptic potential in bullfrog sympathetic ganglia. Brain Res 280:349-354

Akasu T, Ohta Y, Koketsu K (1984) Neuropeptides facilitate the desensitization of nicotinic acetylcholine-receptor in frog skeletal muscle endplate. Brain Res 290:342-347

Akasu T, Tsurusaki M, Ariyoshi M (1986) Presynaptic effects of cholecystokinin octapeptide on neuromuscular transmission in the frog. Neurosci Lett 67:329-333

Ariyoshi M, Tokimasa T, Ohta Y, et al (1984) The effects of histamine on the sensitivity of the frog nicotinic receptor-ionic channel complex. Neurosci Lett 17:S127

Ariyoshi M, Hasuo H, Koketsu K, et al (1985) Histamine is an antagonist of the acetylcholine receptor at the frog end-plate. Br J Pharmacol 85:65-73

Brown DA, Adams PR (1980) Muscarinic suppression of a novel voltage-sensitive K^+ current in a vertebrate neurone. Nature (Lond) 283:673-676

Brown DA, Higgins AJ (1979) Presynaptic effects of γ-aminobutyric acid in isolated rat superior cervical ganglia. Br J Pharmacol 66:108p-109p

Christ DD, Nishi S (1971) Effects of adrenaline on the nerve terminals in the superior cervical ganglion of the rabbit. Br J Pharmacol 41:331-338

Colomo F, Rahamimoff R, Stefani E (1968) An action of 5-hydroxytryptamine on the frog motor end-plate. Eur J Pharmacol 3:272-274

Dun N, Nishi S (1974) Effects of dopamine on the superior cervical ganglion of the rabbit. J Physiol (Lond) 239:155-164

Ewald DA (1976) Potentiation of postjunctional cholinergic sensitivity of rat diaphragm muscle by high-energy-phosphate adenine nucleotides. J Membr Biol 29:47-65

Hasuo H, Akasu T (1986) Luteinizing hormone-releasing hormone inhibits nicotinic transmission in bullfrog sympathetic ganglia. Neurosci Res 3:444-450

Hasuo H, Akasu T (1988a) Presynaptic inhibition of cholinergic transmission by peptidergic neurons in bullfrog sympathetic ganglia. Pflügers Arch 413:206-208

Hasuo H, Akasu T (1988b) Peptidergic inhibition of cholinergic transmission in bullfrog sympathetic ganglia. Jpn J Physiol 38:643-658

Hasuo H, Tokimasa T, Koketsu K (1982) Desensitization of the muscarinic receptor controlling action potential of bullfrog atrial muscles. Jpn J Physiol 32:35-44

Hirai K, Katayama Y (1985) Effect of the endogenous analgesic dipeptide, kyotorphin, on transmitter release in sympathetic ganglia. Br J Pharmacol 85:629-634

Hirai K, Koketsu K (1980) Presynaptic regulation of the release of acetylcholine by 5-hydroxytryptamine. Br J Pharmacol 70:499-500

Jan LY, Jan YN (1982) Peptidergic transmission in sympathetic ganglia of the frog. J Physiol (Lond) 327:219-246

Kaibara K, Akasu T, Tokimasa T, et al (1985) β-Adrenergic modulation of the Na^+-K^+ pump in frog skeletal muscles. Pflügers Arch 405:24-28

Katayama Y, Nishi S (1982) Voltage-clamp analysis of peptidergic slow depolarizations in bullfrog sympathetic ganglion cells. J Physiol (Lond) 333:305-313

Kato E, Kuba K (1980) Inhibition of transmitter release in bullfrog sympathetic ganglia induced by γ-aminobutyric acid. J Physiol (Lond) 298:271-283

Kato E, Kuba K, Koketsu K (1978) Presynaptic inhibition by γ-aminobutyric acid in bullfrog sympathetic ganglion cells. Brain Res 153:398-402

Kato E, Koketsu K, Kuba K, et al. (1985) The mechanism of the inhibitory action of adrenaline on transmitter release in bullfrog sympathetic ganglia: independence of cyclic AMP and calcium ions. Br J Pharmacol 84:435-443

Katz B (1966) Nerve, muscle, and synapse. McGraw-Hill, New York

Kojima-Nishimura M, Akasu T (1983) Substance P facilitates the release of acetylcholine from motor nerve terminals of frogs. Kurume Med J 30:67-71

Kojima M, Akasu T, Koketsu K (1982) Effect of luteinizing hormone-releasing hormone on the release of acetylcholine from motor nerve terminals. Kurume Med J 29:189-192

Koketsu K (1969) Cholinergic synaptic potentials and the underlying ionic mechanisms. Fed Proc 28:101-112

Koketsu K (1971) The electrogenic sodium pump. Adv Biophys 2:77-112

Koketsu K (1974) The mode of actions of transmitter to neurone membrane. Seitaino-Kagaku 25:277-285(in Japanese)

Koketsu K (1981) Electropharmacological actions of catecholamine in sympathetic ganglia. Multiple modes of action to modulate the nicotinic transmission. Jpn J Pharmacol 31(suppl):27P-28P.

Koketsu K (1984) Modulation of receptor sensitivity and action potentials by transmitters in vertebrate neurones. Jpn J Physiol 34:945-960

Koketsu K (1987) Modulation by neurotransmitters of the nicotinic transmission in the vertebrates. In: Dun NJ, Perlman RL (eds) Neurobiology of acetylcholine. Plenum Press, New York, pp 225-238

Koketsu K, Akasu T (1982) Modulation of the slow inward Ca^{2+} current by adrenaline in bullfrog sympathetic ganglion cells. Jpn J Physiol 32:137-140

Koketsu K, Akasu T (1985) Modulation of nicotinic transmission by endogenous substances in sympathetic ganglia. In: Kalsner S (ed) Trends in autonomic pharmacology, vol 3. Taylor & Francis, London, pp 245-256

Koketsu K, Akasu T (1986) Postsynaptic modulation. In: Karczmar AG, Koketsu K, Nishi S (eds) Autonomic and enteric ganglia: transmission and its pharmacology. Plenum Press, New York, pp 273-295

Koketsu K, Minota S (1975) The direct action of adrenaline on the action potentials of bullfrog's (*Rana catesbeiana*) sympathetic ganglion cells. Experientia 31:822-823

Koketsu K, Nishi S (1968) Cholinergic receptors at sympathetic preganglionic nerve terminals. J Physiol (Lond) 196:293-310

Koketsu K, Ohta Y (1976) Acceleration of the electrogenic Na^+ pump by adrenaline in frog skeletal muscle fibres. Life Sci 19:1009-1013

Koketsu K, Yamada M (1982) Presynaptic muscarinic receptors inhibiting active acetylcholine release in the bullfrog sympathetic ganglion. Br J Pharmacol 77:75-82

Koketsu K, Shoji T, Yamamoto K (1974) Effects of GABA on presynaptic nerve terminals in bullfrog (*Rana catesbiana*) sympathetic ganglia. Experientia 30:382-383

Koketsu K, Akasu T, Miyagawa K, et al (1982a) Modulation of nicotinic transmission by biogenic amines in bullfrog sympathetic ganglia. J Auton Nerv Syst 6:47-53

Koketsu K, Miyagawa M, Akasu T (1982b) Catecholamine modulates nicotinic ACh-receptor sensitivity. Brain Res 236:487-491

Koketsu K, Hasuo H, Akasu T (1991) Regulation of acetylcholine release from preganglionic and motor nerve terminals by biogenic substances in vertebrates. In: Feigenbaum J, Hanani M (eds) Presynaptic regulation of neurotransmitter release: a handbook. Freund, London, pp 975-993

Kuba K (1970) Effects of catecholamines on the neuromuscular junction in the rat diaphragm. J Physiol (Lond) 211:551-570

Kuba K, Koketsu K (1975) Direct control of action potentials by acetylcholine in bullfrog sympathetic ganglion cells. Brain Res 89:166-169

Kuba K, Koketsu K (1976) The muscarinic effects of acetylcholine on the action potential of bullfrog sympathetic ganglion cells. Jpn J Physiol 26:703-716

Kuba K, Koketsu K (1978) Synaptic events in sympathetic ganglia. Prog Neurobiol 11:77-169

Kuba K, Kumamoto E (1986) Long-term potentiation of transmitter release induced by adrenaline in bull-frog sympathetic ganglia. J Physiol (Lond) 374:515-530

Kuba K, Tomita T (1971) Noradrenaline action on nerve terminal in the rat diaphragm. J Physiol (Lond) 217:19-31

Kuba K, Kato E, Kumamoto E, et al (1981) Sustained potentiation of transmitter release by adrenaline and dibutyryl cyclic AMP in sympathetic ganglia. Nature (Lond) 291:654-656

Kumamoto E, Kuba K (1983) Independence of presynaptic bimodal actions of adrenaline in sympathetic ganglia. Brain Res 265:344-347

Kumamoto E, Kuba K (1987) Mechanisms regulating the adrenaline-induced long-term potentiation in bullfrog sympathetic ganglia. Pflügers Arch 408:573-577

Minota S, Koketsu K (1977) Effects of adrenaline on the action potential of sympathetic ganglion cells in bullfrogs. Jpn J Physiol 27:353-366

Minota S, Koketsu K (1979) Activation of the electrogenic Na-pump of cardiac muscle fibres by ACh in K-free solutions. Experientia 35:772-773

Miyagawa M, Minota S, Koketsu K (1981) Antidromic inhibition of acetylcholine release from presynaptic nerve terminals in bullfrog's sympathetic ganglia. Brain Res 244:305-313

Morita K, Koketsu K (1979) An analysis of the effect of adrenaline on electrogenic Na^+ pump of visceral nerve fibers in bullfrogs. Jpn J Physiol 29:239-250

Morita K, North RA, Tokimasa T (1982) Muscarinic agonists inactivate potassium conductance of guinea-pig myenteric neurones. J Physiol (Lond) 333:125-139

Morita K, Katayama Y, Koketsu K, et al (1984) Actions of ATP on the soma of bullfrog primary afferent neurons and its modulating action on the GABA-induced response. Brain Res 293:360-363

Nakamura M, Hayashi H, Hirai K, et al (1974) Effects of ATP on sympathetic ganglia from bullfrogs. Jpn J Pharmacol Suppl 24:134

Nishi S (1970) Cholinergic and adrenergic receptors at sympathetic preganglionic nerve terminals. Fed Proc 29:1957-1965

Nishimura T, Tokimasa T, Akasu T (1988a) 5-Hydroxytryptamine inhibits cholinergic transmission through 5-HT$_{1A}$ receptor subtypes in rabbit vesical parasympathetic ganglia. Brain Res 442:399-402

Nishimura T, Tokimasa T, Akasu T, et al (1988b) Presynaptic serotonin (5-HT)-receptor subtypes in rabbit vesical pelvic ganglion. Neurosci Res Suppl 7:177

North RA, Tokimasa T (1983) Depression of calcium-dependent potassium conductance of guinea-pig myenteric neurones by muscarinic agonists. J Physiol (Lond) 342:253-266

Ohta Y, Ariyoshi M, Koketsu K (1984) Histamine as an endogenous antagonist of nicotinic ACh-receptor. Brain Res 306:370-373

Scuka M (1973) Analysis of the effects of histamine on the end-plate potential. Neuropharmacology 12:441-450

Shirasawa Y, Akasu T, Koketsu K (1976) Effects of adrenaline and serotonin on the pump potentials of sympathetic ganglion cell membrane in bullfrogs. J Physiol Soc Jpn 38:270-272

Suetake K, Kojima H, Inanaga K, et al (1981) Catecholamine is released from non-synaptic cell-soma membrane: histochemical evidence in bullfrog sympathetic ganglion cells. Brain Res 205:436-440

Tokimasa T (1984) Muscarinic agonists depress calcium-dependent g_k in bullfrog sympathetic neurons. J Auton Nerv Syst 10:107-116

Tokimasa T, Hasuo H, Koketsu K (1980) Desensitization of the muscarinic receptor of bullfrog atrial muscle. Experientia 36:1200-1201

Tokimasa T, Hasuo H, Koketsu K (1981) Desensitization of the muscarinic acetylcholine receptor of atrium in bullfrogs. Jpn J Physiol 31:83-97

Tsurusaki M (1987) Presynaptic α_2-adrenoceptors mediate inhibition of cholinergic transmission in rabbit vesical parasympathetic ganglia. Kurume Med J 34:213-216

Yamada M, Tokimasa T, Koketsu K (1982) Effects of histamine on acetylcholine release in bullfrog sympathetic ganglia. Eur J Pharmacol 82:15-20

M-Current: From Discovery to Single Channel Currents

D. A. BROWN

Department of Pharmacology, University College London, Gower Street, London, WC1E 6BT, England (UK)

Key words. M-current, Potassium current, Muscarinic receptors, Ganglion, Neuroblastoma, Calcium, Depolarization, Action potentials, Spike adaptation, M-channels, Kv1.2 channels, *eag* channels, Phospholipase C, G proteins.

1 Discovery

This story started with some experiments Andy Constanti and I were doing in 1978 on the depolarizing effects of muscarine on rat sympathetic neurons recorded under 'current-clamp' using a single microelectrode (eventually published in Brown and Constanti 1980). Like others who had investigated the muscarinic 'slow epsp' (e.g., Kobayashi and Libet 1968; Nishi et al. 1968; Weight and Votava 1970; Kuba and Koketsu 1976), we noted that the depolarization was usually acompanied by an apparent increase in 'input resistance' (Fig.1A). However, we also noted some odd features about the response. *First,* the apparent resistance increase was voltage-dependent, that is, it became larger the more the membrane was depolarized from its initial resting potential. Kuba and Koketsu (1976) also noticed this during the slow EPSP in some frog ganglion cells, and suggested that acetylcholine might impair 'delayed rectification'. Although very prescient, at the time this suggestion seemed unlikely to us, at least in the strict sense of the 'delayed rectifier' as that current which repolarizes the action potential, because the voltage-dependence of muscarine's action was quite unlike that of TEA, and muscarine did not lengthen the action potential. *Second,* muscarine produced a most remarkable increase in excitability: when we injected second-long depolarizing currents, instead of a single action potential, we generated a sustained train of spikes; i.e., the cell switched behavior from phasic to tonic firing (Fig1A).

It was difficult to see how we could explore this further in the rat ganglion, using such relatively primitive recording, but an opportunity came to do so in Paul Adams' lab in Galveston, using the two-electrode voltage-clamp on the larger, adendritic bullfrog sympathetic neurons (Brown and Adams 1980; Adams et al.

1982a, b). For the initial experiments we used the voltage-clamp equivalent of the 'conductance testing' protocol Constanti and I had used for the rat ganglion experiments; that is, we held the cell at a relatively depolarized level (-40mV or above), where we knew that muscarine would produce a large response, and then gave 0.5- to 1-s hyperpolarizing commands at intervals of 5 s or so, to measure 'input conductance'. Muscarine produced the anticipated inward (depolarizing) current and duly reduced the 'input conductance'. However, the most striking feature was the nature of the current response to the hyperpolarizing step – a slow inward current relaxation, followed by an equally slow outward relaxation on repolarizing the cell – and the fact that muscarine selectively reduced, and sometimes abolished, these slow relaxations (Fig.1B).

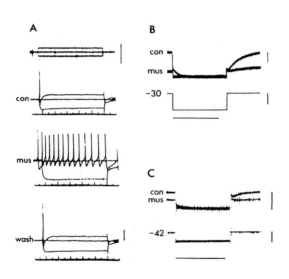

Fig.1A-C. Reproductions of original oscillograph/oscilloscope records about 1978-1980. Calibration bars: current, 2 nA; voltage, 20mV. **A** Response of an isolated rat superior cervical sympathetic ganglion to 10 µM muscarine (mus) recorded with a single microelectrode. (Adapted from Brown and Constanti 1980; q.v. for experimental details.) Records show voltage responses to 1-s injections of ± 0.4 nA current (top trace) before ('con'), during and after ('wash') addition of muscarine. Note that muscarine depolarized the cell (voltage trace shift from time-baseline) and induced repetitive firing. **B** Superimposed voltage-clamp records of current deflections produced by a 30-mV, 600-ms hyperpolarizing step from a holding potential of –30 mV before ('con') and during ('mus') application of 10 µM muscarine to a bullfrog ganglion cell. Twin microelectrode recording. (Experiment with P. R. Adams; adapted from Brown 1983.) **C** Single microelectrode 'switch-clamp' recording from a neurone in an isolated rat superior cervical ganglion using a similar protocol, at a holding potential of –42 mV. (Experiment with A. Constanti; adapted from Brown 1983).

These relaxations turned out to result from the deactivation and reactivation, respectively, of a time- and voltage-dependent K^+ current distinct from the delayed rectifier current. Thus, it had a low threshold (~ -60 mV), activated slowly (with a time-constant down to ~150 ms) and, in an apparently exponential manner without the sigmoidal onset of the delayed rectifier, did not inactivate over minutes or even hours, and was relatively insensitive to TEA. [Channel permeation properties also differ from those of delayed rectifier channels: M-channels show negligible permeability to Na^+ and do not exhibit multi-ion pore behaviour (Cloues and Marrion, 1996; Block and Jones 1996)]. Initial analysis suggested that gating followed very simple Boltzmann kinetics, describable in terms of two channel conformations (closed and open), whose transition was governed by a single tetravalent voltagesensor, although subsequent work suggests this to be much oversimplified (see, e.g., Owen et al. 1990; Marrion et al. 1992). It was termed an 'M-current' because it was inhibited by muscarinic agonists (and partly for want of a better term): this is not particularly definitive, because other currents are inhibited by muscarinic agonists, even in sympathetic neurons (e.g., Cassell and McLachlan 1987), and because muscarine does not always inhibit the M-current (e.g., Jones 1987); the cell has to have the right muscarinic receptor (m1 or m3: Fukuda et al. 1988) to make the connexion.

2 Distribution

Constanti and I confirmed that the same current was indeed present in rat sympathetic neurons (Constanti and Brown 1981: see Fig. 1C). They were subsequently identified in a variety of other neurons including some central neurons such as hippocampal and cortical pyramidal cells (including human neurons) (see Brown 1988b, for review; McCormick and Williamson 1989), and in some nonneural cells, such as frog smooth muscle cells (Sims et al. 1985). Currents resembling M-currents in one or more aspects ('M-like currents') have also been recorded from frog retinal ganglion cells (I_{Kx}: Beech and Barnes 1989) and from some neural cell lines such as NG108-15 neuroblastoma x glioma hybrid cells (Higashida and Brown, 1986; Brown and Higashida, 1988a; Robbins et all 1992) and PC12 cells (Villarroel et al. 1989). [I call these 'M-like' because, while showing rather similar kinetics, they differ from ganglionic M-currents in certain other respects. For example, I_{Kx} and $I_{K(M,ng)}$ show a different sensitivity to, and blocking mechanism of, divalent cations (Robbins et al.1992; Wollmuth 1994). $I_{K(M,ng)}$ also differs from the ganglionic M-current in its sensitivity to some organic K^+-channel blockers (Robbins et al. 1992); for instance, it is some 20 times less sensitive to linopirdine (A.K.Filippov, unpublished; cf. Lamas et al. 1997). These differences might be of significance in deducing which molecular species of K^+-channel(s) generate the M-current (see below).]

3 What the M-Current Does: Physiological Function

When a healthy ganglion cell is depolarized from its rest potential of about −60 mV to −30 mV, the membrane slope conductance increases by a factor of at least 10 times as the M-current is activated. This provides a strongly-stabilizing, negative-feedback influence on membrane potential perturbation: in a sense, it forms the cell's own 'voltage-clamp' device, albeit a rather slow one compared with an electronic clamp (see Fig.6 in Brown 1988a).

What about more transient depolarizations, such as synaptic and action potentials? Both experimental (Tosaka et al. 1983) and computational (Brown 1988a) tests show that $I_{K(M)}$ is activated by excitatory synaptic potentials, accelerating their decline and inducing an undershoot. Although too slow to activate significantly during a single action potential (Lancaster and Pennefather 1987), $I_{K(M)}$ activates strongly during repetitive spike discharges: as well as producing a gradual hyperpolarization, this raises the threshold for spike generation and so arrests the spike train (Brown 1983; Jones and Adams 1987). However, action potentials also increase Ca^{2+} influx, which in turn activates the slow Ca^{2+}-dependent K^+ current ('AHP' current; Pennefather et ai. 1985). This induces an additional 'braking' effect on spike discharges. Thus, $I_{K(M)}$ and I_{AHP} seem to act synergistically in a 'belt-and-braces' manner: if one is inhibited, the other is consequently activated more strongly, and both have to be inhibited for the full discharge to be expressed (Jones and Adams 1987; Yamada et al. 1989; cf. Brown 1988a). Nevertheless, $I_{K(M)}$ plays a unique role in the overall physiology of autonomic function, since it appears to be the primary factor in determining whether the discharge of a sympathetic ganglion cell is 'phasic' or 'tonic' in nature (Cassell et al. 1986; Wang and McKinnon 1995).

Is the M-current active at rest? Initial modelling in frog ganglion cells (Adams et al. 1982a) suggested yes, but this was in cells subject to a depolarizing current (via the microelectrode implements): more careful studies (Tosaka et al. 1983; Jones 1989) indicated that there is negligible activation at the normal resting potential. The same seems to be true for the M-like current in NG108-15 cells (Robbins et al. 1992). In rat sympathetic neurons 'minimally disturbed' by perforated patch recording, with a resting membrane potential of about -60 mV at $20°$ $-23C$ (-58.3 ± 7.3 mV: Selyanko et al. 1992), M-channels are just teetering on the threshold: openings in cell-attached recordings are very sparse ($P_o < 0.01$), and the M-channel blocker, linopirdine, produced depolarization of a few millivolts in about half the cells held at −60 mV (Lamas et al. 1997). What actually sets the resting potential in cells expressing M-currents is still not all that clear (see Jones 1989).

4 M-Channels

Single M-channel and M-like-channel currents have now been recorded from frog (Marrion, 1993, 1996), rat (Owen et al. 1990; Selyanko et al. 1992; Stansfeld et al. 1993; Selyanko and Brown, 1993, 1996), and NG108-15 (Selyanko et al. 1995) cells. Notwithstanding some initial contradictions, there is now reasonable convergence of opinion regarding some of their basic properties (see Marrion, 1997). They show two main conductance levels, of around 7 and 11 pS in rat ganglia (cell-attached, asymmetric [K^+]), and 10 and 15 pS in frog ganglia (cell-attached, symmetrical [K^+]). [Direct comparison is difficult because single channel currents rectify, so it is more useful to quote permeabilities: the permeability of the "11 pS" channel in excised patches from rat neurons is ~5 x 10^{-14} cm^2s^{-1}]. Also, both frog and rat channels show two time constants in the steady-state open-time distributions, implying that the original two-state kinetic description based on macroscopic currents (Adams et al. 1982a) is indeed an oversimplification. Steady-state kinetics in frog cells has been fitted with a model in which the channels switch between two 'modes' of gating (high and low P_o) – a simple C-O gating for mode 1 (low P_o) and C-C-O gating for mode 2 (high P_o) (Marrion, 1992). No evidence for such modal gating could be deduced from stability plots of rat channel behaviour (Selyanko and Brown 1996a, and unpublished); instead, both steady-state open/shut time distributions and ensemble kinetics during a depolarizing step could best be described by a linear model, C_L-O_S-C_M-O_L-C_S, where L,M and S are long, medium, and short dwell-times (Selyanko and Brown 1997 and unpublished). Of course, these are only operational schemes: their physical significance is totally unclear as yet.

5 M-Channel Clones?

Thus far, two cloned K^+ channel proteins have been suggested as possible components of M-channels, the *Shaker* homologue Kv1.2 (RAK; Morielli and Peralta 1995), and rat *ether-a-go-go* (r-*eag;* Stansfeld et al. 1997). Although both show some "M-like" features, neither can be regarded as the true "M-channel" as such, at least, not in homomeric form.

5.1 Kv1.2

In a search for the M-like channel in NG108-15 cells, Yokoyama and colleagues (Yokoyama et al. 1989) isolated two clones, the *Shaker* homolog NGK1 (Kv1.2) and the *Shaw* homolog NGK2 (Kv3.1, which probably underlies the inactivating delayed rectifier current recorded from these cells: Robbins and Sim 1990). Expressed Kv1.2 currents show some rather obvious differences from the M-like current; for example, it normally activates rather rapidly and (perhaps most strikingly) is potently blocked by dendrotoxins (Werkmann et al. 1992; even in heteromultimers: Hopkins 1997), whereas the M-like channel is not (Selyanko et al. 1995). On the other hand, there are also some similarities: expressed currents have a fairly low threshold (~-40 mV; Ito et al. 1992), are inhibited by stimulating m1 mAChRs (Huang et al. 1993), and show a comparable conductance (~11 pS) and kinetic appearance at the single channel level (Yokoyama et al. 1993). Further, under certain circumstances, ensemble channel currents can exhibit time-dependent activation quite reminiscent of M-like channels (A.A.Selyanko and

(Delmas et al. 1998a,b; Haleyet al. 1997). (d) On the other hand, M-current was totally abolished by overexpressing a GTPase-deficient mutant of $G\alpha_q$ (and $G\alpha_{11}$, so $G\alpha_{11}$ is *potentially* capable of transducing an M-current inhibiting signal: Haley et al. 1997). Then the cell shows the sort of repetitive discharges we originally saw on applying muscarine (except even more vigorous; Fig. 2).

6.3 "Messengers"

Ganglionic M-channels, and NG108-15 M-like channels, when recorded in cell-attached patches are closed by stimulating receptors outside the patch (Selyanko et al. 1992; Marrion 1993; Selyanko et al. 1995), but not detectably with the agonist inside the patch electrode (Selyanko et al. 1992). So something must carry the "message" from the extrapatch receptors to the intrapatch channels, either through the membrane or via the cytosol. The idea of a remote, possibly multistep, linkage is also supported by the slow onset of inhibition (Jones 1991); for example, even after synaptically-released acetylcholine has "hit" the nicotinic receptors, there is a further delay of ~200-300 ms at 34°C before any current resulting from M-channel inhibition can be detected, and this lengthens to 1-2 s at a low (24°C) temperature (Brown et al. 1995).

There is no lack of candidates for this messenger role. Marrion (1997b) has assessed their various credentials in detail, and one can only conclude that the position is still vacant (at least, for the muscarinic messenger). Thus, one of the more obvious is Ca^{2+} ions. It is clear that M-channels can be closed by raising intracellular $[Ca^{2+}]$ – either directly (in rats; Selyanko and Brown, 1996a,b) or indirectly via calcineurin (in frogs; Marrion 1996) – and it seems likely that the channels may be under some tonic inhibition at resting levels of intracellular $[Ca^{2+}]$ (Selyanko and Brown, 1996b), but there is still no conclusive evidence to show that Ca^{2+} is responsible for the inhibition produced by muscarinic agonists. However, Cruzblanca et al. (1997) have recently obtained better evidence to suggest that Ca^{2+} might be responsible for the inhibition of the rat ganglion M-current by bradykinin (which is also mediated by G_q or G_{11}; Jones et al. 1995), so the "Ca^{2+} story" may not be over yet!

Acknowledgements. I thank all my colleagues named above for their contributions to this saga. The experiments were mostly theirs; the views are my own. I also apologize to those whose names are omitted largely for reasons of space.

References

Adams PR, Brown DA, Constanti A (1982a) M-currents and other potassium currents in bullfrog sympathetic neurones. J Physiol 330:537-572

Adams PR, Brown DA, Constanti A (1982b) Pharmacological inhibition of the M-current. J Physiol 332:223-262

Beech DJ, Barnes S (1989) Characterization of a voltage-gated K^+ channel that accelerates the rod response to dim light. Neuron 3:573-581

Bernheim L, Mathie A, Hille B (1992) Characterization of muscarinic receptor subtypes inhibiting Ca^{2+} and M current in rat sympathetic neurons. Proc Natl Acad Sci USA 89:9544-9548

Block B, Jones SW (1966) Ion permeation and block of M-type and delayed rectifier potassium channels. J Gen Physiol 107:473-488

Brown DA (1983) Slow cholinergic excitation – a mechanism for increasing neuronal excitability. Trends Neurosci 6:302-307

Brown, DA (1988a) M currents. In: Narahashi, T (ed) Ion channels,vol1. Plenum, New York, pp 55-99

Brown DA (1988b) M-currents: an update. Trends Neurosci 11:294-299

Brown DA, Adams PR (1980) Muscarinic suppression of a novel voltage-sensitive K^+ current in a vertebrate neurone. Nature 283:673-676

Brown DA, Buckley NJ, Caulfield MP, et al (1995) Coupling of muscarinic acetylcholine receptors (mAChRs) to neural ion channels: closure of some K^+ channels. In: Wess J (ed) Molecular mechanisms of muscarinic acetylcholine receptor function. Landes, New York, pp 165-181

Brown DA, Constanti A (1980) Intracellular observations on the effects of some muscarinic agonists on rat sympathetic neurones. Br J Pharmacol 70:593-608

Brown DA, Higashida H (1988) Voltage- and calcium-activated potassium currents in mouse neuroblastoma x rat glioma hybrid cells. J Physiol 397: 149-165

Cassell JF, Clark AL, McLachlan EM (1986) Characteristics of phasic and tonic sympathetic ganglion cells of the guinea-pig. J Physiol 372:457-483

Cassell JF, McLachlan EM (1987) Muscarinic agonists block five different potassium conductances in guinea-pig sympathetic neurones. Br.J.Pharmacol 91:259-261

Caulfield MP, Jones S, Vallis Y, et al (1994) Muscarinic M-current inhibition via $G\alpha_{q/11}$ and α-adrenoceptor inhibition of Ca^{2+} current via $G\alpha_o$ in rat sympathetic neurones. J Physiol 477:415-422

Cloues R, Marrion NV (1996) Conduction properties of the M-channel in rat sympathetic neurons. Biophys J 70:806-812

Constanti A, Brown DA (1981) M-currents in voltage-clamped mammalian sympathetic neurones. Neurosci Lett 24:289-294

Cruzblanca H, Koh D-S, Hille B (1997) Bradykinin inhibits the M-current through a Ca^{2+}-dependent pathway in sympathetic neurons. Soc Neurosci Abstr 23:1743

Delmas P, Brown DA, Dayrell M, et al (1998a) On the role of endogenous G-protein $\beta\gamma$ subunits in I_{Ca} inhinition by neurotransmitters in rat sympathetic neurones. J Physiol 506:319-329

Delmas P, Abogadie FC, Dayrell M, et al (1998b) G-proteins and G-protein subunits mediating cholinergic inhibition of N-type calcium currents. Eur J Neurosci (in press)

Filippov AK, Webb TE, Barnard EA, et al (1998) P2Y2 nucleotide receptors heterologously-expressed in rat sympathetic neurones inhibit both Ca^{2+} - and K^+-currents via different G proteins. J Physiol (in press)

Fukuda K, Higashida H, Kubo T, et al (1988) Selective coupling with K^+ currents of muscarinic acetylcholine receptor subtypes in NG108-15 cells Nature 335:355-358

Haley JE, Delmas P, Abogadie, FC, et al (1997) Muscarinic inhibition of the M-current is mediated by the α subunit of Gq. J Physiol 504P:176-177P

Hamilton SE, Loose MD, Qi M, et al (1997) Disruption of the m1 receptor gene ablates muscarinic receptor-dependent M current regulation and seizure activity in mice. Proc Natl Acad Sci USA 94:13311-13316

Higashida H, Brown DA (1986) Two polyphosphatidylinositide metabolites control two K^+ currents in a neuronal cell line. Nature 323:333-335

Hille, B (1994) Modulation of ion-channel function by G protein-coupled receptors. Trends Neurosc 17:531-536

Hopkins WE (1997) Specificity of toxin inhibition of heteromultimeric potassium channels expressed in *Xenopus* oocytes. Soc Neurosci Abstr 23:1746

Huang X-Y, Morielli AD, Peralta EG (1993) Tyrosine kinase-dependent suppression of a potassium channel by the G protein-coupled m1 muscarinic acetylcholine receptor. Cell 75:1145-1156

Ikeda SR, Lovinger DM, McCool BA, et al (1995) Heterologous expression of metabotropic glutamate receptors in adult rat sympathetic neurons: subtype-specific coupling to ion channels. Neuron 14:1029-1038

Ito Y, Yokoyama S, Higashida H (1992) Potassium channels cloned from neuroblastoma cells display slowly inactivating outward currents in *Xenopus* oocytes. Proc R Soc Lond Ser B 248:95-101

Jones S, Brown DA, Milligan G, et al (1995) Bradykinin excites rat sympathetic neurons by inhibition of M current through a mechanism involving B_2 receptors and $G_{\alpha q/11}$. Neuron 14:399-405

Jones SW (1987) A muscarine-resistant M-current in C cells of bullfrog sympathetic ganglia. Neurosci Lett 74:309-314

Jones SW (1989) On the resting potential of isolated frog sympathetic neurons. Neuron 3:153-161

Jones SW (1991) Time course of receptor-channel coupling in frog sympathetic neurons. Biophys J 60:502-507

Jones SW, Adams PR (1987) The M-current and other potassium currents of vertebrate neurons. In: Kaczmarek K & Levitan I B (eds) Neuromodulation. Oxford University Press, New York, pp159-186

Kobayashi H, Libet B (1968) Generation of slow post-synaptic potentials without increase in ionic conductance. Proc Natl Acad Sci USA 60:1304-1311

Koch WJ, Hawes BE, Inglese J, et al (1994) Cellular expression of the carboxyl terminus of a G protein-coupled receptor kinase attenuates Gβ-mediated signaling. J Biol Chem 269:6193-6197

Kuba K, Koketsu K (1976) Analysis of the slow excitatory post-synaptic potential in bullfrog sympathetic ganglion cells. Jpn J Physiol 26:651-669

Lamas JA, Selyanko AA, Brown DA (1997) Effects of a cognition-enhancer, linopirdine (DuP 996) on M-type potasium currents and some other voltage- and ligand-gated membrane currents in rat sympathetic neurons Eur J Neurosci 9:605-616

Lancaster B, Pennefather P (1987) Potassium currents evoked by brief depolarizations in bull-frog sympathetic ganglion cells J Physiol 387:519-548

McCormick DA, Williamson A (1989) Convergence and divergence of neurotransmitter action in human cerebral cortex. Proc Natl Acad Sci USA 86:8098-8102

Marrion NV (1993) Selective reduction of one mode of M-channel gating by muscarine in sympathetic neurons. Neuron 11:77-84

Marrion NV (1996) Calcineurin regulates M channel modal gating in sympathetic neurons. Neuron 16:163-173

Marrion NV (1997a) Does r-EAG contribute to the M-current? Trends Neurosci 20:243

Marrion NV (1997b) Control of M-current. Physiol Rev 59:483-504

Marrion NV, Adams PR, Gruner W (1992) Multiple kinetic states underlying macroscopic M-currents in bullfrog sympathetic neurons. Proc R Soc Lond Ser B 248:207-214

Marrion NV, Smart TG, Marsh SJ, et al (1989) Muscarinic suppression of the M-current in the rat sympathetic ganglion is mediated by receptors of the M1-subtype. Br J Pharmacol 98:557-573

Morielli AD, Peralta EG (1995) Suppression of a potassium channel by G-protein coupled receptors. Life Sci 56:1035

Nishi S, Soeda H, Koketsu K (1968) Unusual nature of ganglionic slow e.p.s.p. studied by a voltage-clamp method. Life Sci 8:33-42

Owen DG, Marsh SJ, Brown DA (1990) M-current noise and putative M-channels in cultured rat sympathetic ganglion cells. J Physiol 431:269-290

Pennefather P, Lancaster B, Adams PR, et al (1985) Two distinct Ca^{2+}-dependent K^+ currents in bullfrog sympathetic ganglion cells. Proc Natl Acad Sci USA 82: 3040-3044

Robbins J, Sim J (1990) A transient outward current in NG108-15 neuroblastoma x glioma hybrid cells. Pflueg Arch 416:130-137

Robbins J, Caulfield MP, Higahsida H, et al (1991) Genotypic m3-muscarinic receptors preferentially inhibit M-currents in DNA-transfected NG108-15 neuroblastoma x glioma hybrid cells. Eur J Neurosci 3:820-824

Robbins J, Trouslard J, Marsh S, et al (1992) Kinetic and pharmacological properties of the M-current in rodent neuroblastoma x glioma hybrid cells. J Physiol 451:159-185

Robbins J, Marsh SJ, Brown DA (1993) On the mechanism of M-current inhibition by muscarinic m1 receptors in DNA-transfected NG108-15 cells. J Physiol 469:153-178

Selyanko AA, Brown DA (1993) Effects of membrane potential and muscarine on potassium M-channel kinetics in rat sympathetic neurone. J Physiol 472:134-139

Selyanko AA, Brown DA (1996a) Intracellular calcium directly inhibits potassium M-channels in excised patches from rat sympathetic neurons. Neuron 16:151-162

Selyanko AA, Brown DA (1996b) Regulation of M-type potassium channels in mammalian sympathetic neurons: action of intracellular calcium on single channel currents. Neuropharmacology 35:933-947

Selyanko AA, Brown DA (1997) Gating kinetics and simulation of potassium M-channels in rat sympathetic neurons. Soc Neurosci Abstr 27:1194

Selyanko AA, Robbins J, Brown DA (1995) Putative M-type potassium channels in neuroblastoma glioma hybrid cells: inhibition by muscarine and bradykinin. Receptors Channels 3:145-159

Selyanko AA, Stansfeld CE, Brown DA (1992) Closure of potassium M-channels by muscarinic acetylcholine-receptor stimulants requires a diffusible messenger. Proc R Soc Lond Ser B 250:119-125

Sims SM, Singer JJ, Walsh JV Jr (1985) Cholinergic agonists suppress a potassium current in freshly dissociated smooth muscle cells of the toad. J Physiol 367:503-529

Stansfeld CE, Marsh SJ, Gibb AJ, et al (1993) Identification of M-channels in outside-out patches excised from sympathetic ganglion cells Neuron 10:639-654

Stansfeld CE, Roeper J, Ludwig J, et al (1996) Elevation of intracellular calcium by muscarinic receptor activation induced a block of voltage-activated rat *ether-a-go-go* channels in a stably transfected cell line. Proc Natl Acad Sci USA 93:9910-9914

Stansfeld CE, Ludwig J, Roeper J, et al (1997a) A physiological role for the *ether-a-go-go* K^+ channels? Trends Neurosci 20:13-14

Stansfeld CE, Ludwig J, Roeper J, et al (1997b) Does r-EAG contribute to the M-current? Reply. Trends Neurosci 20:243-244

Terlau H, Ludwig J, Steffan R, et al (1996) Extracellular Mg^{2+} regulates activation of rat *eag* potassium channel. Pflueg Arch 432:301-312

Tosaka T, Tasaka J, Miyazaki T, et al (1983) Hyperpolarization following activation of K^+ channels by excitatory postsynaptic potentials. Nature 305:148-150

Villaroell A, Marrion NV, Lopez H, et al (1989) Bradykinin inhibits a potassium M-like current in rat phaeochromocytoma PC12 cells. FEBS Lett 255:42-46

Wang H-S, McKinnon D (1995) Potassium currents in rat prevertebral and paravertebral sympathetic neurones: control of firing properties. J Physiol 485:319-335

Weight FF, Votava Z (1970) Slow synaptic excitation in sympathetic ganglion cells: evidence for synaptic inactivation of potassium conductance. Science 170:755-758

Werkman TR, Kawamura T, Yokoyama S et al (1992) Charybdotoxin, dendrotoxin and mast cell degranulating peptide block the voltage-activated K^+ current of fibroblast cells stably transfected with the NGK1 (Kv1.2) K^+ channel complementary DNA. Neurosciences 50:935-946

Wollmuth LP (1994) Mechanism of Ba^{2+} block of M-like L channels of rod photoreceptors of tiger salamanders. J Gen Physiol 103: 45-66

Yamada WM, Koch C, Adams PR (1989). Multiple channels and calcium dynamics. In: Koch C & Segev I (eds) Methods in neuronal modelling. Bradford, Cambridge MA, pp 97-133.

Yokoyama S, Imoto K, Kawamura H, et al (1989) Potassium channels from NG108-15 neuroblastoma-glioma hybrid cells: primary structure and functional expression from cDNAs. FEBS Lett 259:37-42

Yokoyama S, Kawamura T, Ito Y, et al (1993). Potassium channels cloned from NG108-15 neuroblastoma-glioma hybrid cells. Ann N Y Acad Sci 707:60-73

Note added in proof: Wang et al (1998: Science, 282, 1890-1893) have provided evidence to suggest that M channel in rat sympathetic neurons is composed of a heteromeric assembly of *KCNQ2* and *KCNQ3* gene products. The M-like current in NG108-15 cells is probably a mixed current, partly generated by *KCNQ2/3* channels, and partly by *merg1* channels (A.A. Selyanko et al., unpublished).

Properties of Muscarine-Sensitive Potassium Currents in Vertebrate Nerve Cells

T. TOKIMASA and T. NISHIMURA

Department of Physiology Tokai University School of Medicine
Bohseidai, Isehara 259-1193 Japan

Key Words. Potassium current; Acetylcholine; Muscarinic receptors; M-current;
K-creep; Intracellular calcium; Calcium transient; fura-2; Myosin light chain;
Myosin light chain kinase; Calcineurin; Calmodulin; Sympathetic neurons;
Myenteric neurons; Submucous neurons

Summary. Two problematic issues for potassium channel currents underlying
slow EPSPs have been reviewed. It is clear that M-current does not require a
priming calcium current for its activation, but intracellular concentration of
calcium ions does control the amplitude of M-current by shifting its activation
curve along the voltage axis. This implies that M-current can be recruited by
calcium ions to be inhibited by ACh under certain circumstances. It is also clear
that K-creep is a potassium current through channels that can be opened by
intracellular calcium ions. Unlike M-current, K-creep is already activated at rest.
The kinetic model predicts that the rate constant for dissociation of the species
CaX (Goldstein et al. 1974) might differ significantly among cell types, and that
this rate constant might be a primary target for the transmitter actions.

Acetylcholine (ACh) acting at muscarinic receptors often mediates slow excitatory
postsynaptic potentials (EPSPs) in both the peripheral and central nervous system.
Mechanisms for these slow EPSPs have been best studied in autonomic ganglia,
and it has been concluded that ACh does not produce these EPSPs by acting
directly to open cation channels; instead, it acts indirectly via guanosine
nucleotides binding proteins to close potassium channels (Kuba and Koketsu
1978; Brown 1988; Mihara 1993; Smith 1994; Akasu and Nishimura 1995).

At least two types of potassium channel currents have been reported that can be
inhibited by synaptically released ACh. One is a noninactivating M-type current
originally described in bullfrog sympathetic ganglion cells (Adams and Brown
1982; Adams et al 1982a; Brown and Adams 1980). The other is K-creep
originally defined in guineapig enteric neurons as a persistent calcium-sensitive

potassium current contributing to the resting membrane potential (Morita et al. 1982; North and Tokimasa 1983; 1987; Akasu and Tokimasa 1989).

The main purpose of the present study was to review problematic issues for these currents (one issue for each) since both issues are central to our understanding of how M-current and K-creep function physiologically and hence how ACh can polarize the postsynaptic membrane for tens of seconds or even minutes by inhibiting either of these currents. For the purpose of discussion, profiles of M-current and K-creep are described briefly in the first section, followed by the calcium-dependence of M-current and the kinetic model for K-creep (and its modulation by transmitters) in the second and third section, respectively.

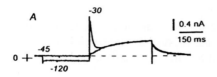

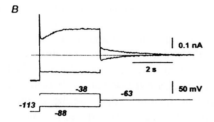

Fig. 1A, B. Profiles of M-current and K-creep. **A** Results (modified from Tokimasa and Nishimura 1997) were obtained in the whole-cell configuration from a dissociated bullfrog sympathetic neuron superfused with Ringer solution containing TEA (30 mM). The cell membrane potential was held at -45 mV, then stepped to -30 mV either directly or following the prepulse to -120 mV. Superimposed current traces (voltage traces not shown) at -30 mV represent M-current (records without prepulse), and A-current followed by M-current. **B** Superimposed current and voltage traces were obtained from an intact submucous neuron of the guinea-pig cecum using single-electrode/voltage-clamp tech-nique. Step depolarizations from -113 to -63 mV were interrupted by the second step to either -88 mV (records without A-current) or -38 mV. An outwardly developing current on depolarization from -88 to -63 mV represents the activating K-creep at -63 mV, indicating the presence of K-creep at rest.

1 Profiles of M-Current and K-Creep

Properties of these currents were studied in intact ganglion cells with single- or two-electrode/voltage-clamp technique, but recent studies for M-current involve whole-cell/voltage-clamp experiments in dissociated (or cultured) ganglion cells (Smith 1994; Tokimasa and Akasu 1995; Marrion 1997).

Figure 1A shows an example of our records obtained from a dissociated bullfrog sympathetic neuron; the cell membrane potential was held at -45 mV, then stepped to -30 mV for 400 ms either directly or following a 300-ms hyperpolarizing prepulse to -120 mV. A difference between the current responses with and without the prepulse yielded the rapidly inactivating potassium current usually referred to as A-current (Adams et al. 1982a; Tokimasa et al 1991) at -30 mV in the absence of M-current. Figure 1B shows a mixture of A-current and K-creep obtained from a submucous neuron of the guineapig cecum. Owing to the absence of inactivation, both M-current and K-creep function as 'inherent break' in unclamped cells (Tokimasa and Akasu 1995).

The M-current changes its amplitude single exponentially with time (Adams et al. 1982a). The time constant ranges from a few milliseconds to some 150 ms depending on the membrane potential at which it is measured. Time course of activation and deactivation of K-creep is also a single exponential function of time (North and Tokimasa 1987). The time constant of K-creep is 20-30 times greater than that of M-current when measured at the same voltage and temperature.

Barium (0.1-1 mM) blocks M-current without significantly affecting the delayed rectifier-type potassium current and A-current; tetraethylammonium (TEA)(up to 30 mM) and 4-aminopyiridine (4-AP)(up to 2 mM) are without significant effects on M-current (Adams et al. 1982b; Tokimasa et al. 1991; Tokimasa and Nishimura 1997). Pharmacology of K-creep is different from that of M-current in one respect (North and Tokimasa 1984, 1987). TEA (30 mM) inhibits K-creep by more than 85% (North and Tokimasa 1987). Apamin is without effects on both currents (Tokimasa and Akasu 1995).

2 Calcium Dependence of M-Current

Adams et al. (1982a) described that M-current was not a class of calcium-activated potassium current in a sense that it did not seem to require an influx of calcium ions for its activation. Tokimasa (1985) proposed that M-current can be inhibited (rather than activated) by massive influx of calcium ions during the calcium spike. Marrion et al. (1991) proposed that intracellular calcium ions have both facilitatory and inhibitory actions on M-current.

We have recently demonstrated that M-current can be activated, or more precisely potentiated, following an inward calcium current (Figure 2) (Hua et al. 1994; Tokimasa 1996; Tokimasa et al. 1996a; 1997). Ratiometric measurement of the fura-2 fluorescence demonstrated that the intracellular concentration of calcium ions ($[Ca]_i$) was increased to about 1 µM during the M-current potentiation (Tokimasa et al. 1997). The steady-state activation curve of M-current was examined with the pipette solution having different free calcium concentrations (Tokimasa et al. 1996a,b). With free calcium at 10 nM, M-conductance was the half-maximal at about -20 mV with the slope factor of 9.6 mV. Respective values were about -42 mV and 8 mV with free calcium at 1 µM. These results indicated that the fraction of M-conductance activated at -65 mV increased from less than 1% to more than 5% in a calcium-dependent manner. Although direct evidence is still lacking, an influx of calcium ions through nicotinic receptors (Tokimasa and North 1984; Trousland et al. 1993) may very well result in an elevation of [Ca]$_i$ and subsequent 'recruitment' of M-conductance, which is normally not available for the muscarinic actions of ACh.

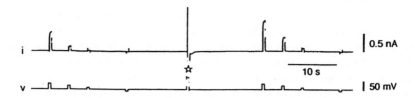

Fig. 2. Calcium-dependent potentiation of M-current. Continuous recordings on chart paper. Holding potential was -65 mV. Middle: calcium current (asterisk, inward and outward peaks off scale) activated during a 0.5-s depolarizing command pulse to 0 mV. Left (before calcium current) and right (after calcium current), relaxations on the current trace in response to command pulses to -35, -45, -55, and -75 mV for 0.5 s. (Modified from Tokimasa et al. 1996a)

Biochemical reactions responsible for the calcium-dependent potentiation of M-current are not yet fully understood (Marrion 1997; Tokimasa and Akasu 1995). Several lines of evidence strongly suggest that calcium/calmodulin-dependent phosphorylation of the 20-kDa light chain of myosin by myosin light chain kinase (MLCK) could be involved (Akasu et al. 1993; Tokimasa et al. 1995). Consistent with this, Pfaffinger (1988) reported that M-current was washed out unless the pipette solution contained a hydrolyzable form of ATP (1-5 mM). Regarding dephosphorylation of myosin or otherwise closely related proteins, Marrion (1996) reported that M-current can be inhibited by a calcium/calmodulin-dependent phosphatase known as calcineurin.

3 Kinetic Model for K-Creep

Myenteric neurons of the guinea pig ileum have been classified using the criteria of Nishi and North (1973) and Hirst et al. (1974). Neurons responding to focal stimulation of nerve strands entering the ganglion with the fast EPSP (nicotinic in nature) have been called S/type-1 cells. Those which show a prominent calcium-dependent afterhyperpolarization following a single action potential have been called AH/type-2 cells.

K-Creep contributes to about 20% of the resting membrane conductance in AH/type-2 cells; 4 nS at -60 mV on average in absolute term (North and Tokimasa 1983, 1987). It is likely that a tonic influx of calcium ions, which might be a counterpart of a 10- to 20-pA persistent inward calcium current at -60 mV (Akasu and Tokimasa 1989), occurs in somewhat leaky manner to maintain the resting level of [Ca]$_i$ at 110-120 nM (Tatsumi et al. 1988) and hence to hold K-creep at 4 nS. The afterhyperpolarization of the action potential was shown to result from an influx of calcium ions during the spike and subsequent activation of K-creep.

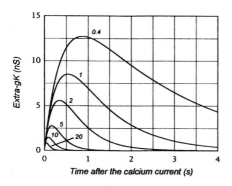

Fig. 3. Kinetic model for K-creep. This graph shows the time course with which the K-creep increases then returns to control after the termination of the inward calcium current. Since K-creep at rest varies with K_{-1} (e.g., 4 nS with K_{-1} at 0.4 s^{-1}), the ordinate denotes an extra K-creep (in nS) recruited by the calcium current. (Modified from Tokimasa and Akasu 1995)

The plateau principle (Goldstein et al. 1974) has been applied to model the kinetics of K-creep in that the second compartment simply represents the species CaX in a reaction of the form

$$Ca + X \overset{K_{+1}}{\underset{K_{-1}}{\rightleftharpoons}} CaX,$$

where Ca represents the free intracellular calcium, while K_{+1} (M^{-1} s^{-1}) and K_{-1} (s^{-1}) denote the rate constant for formation of CaX and dissociation of CaX (North and Tokimasa 1987; Tokimasa and Akasu 1995). If the concentration of CaX ([CaX]) linearly holds K-creep, then fraction of K-creep at rest would be given by [Ca]/([Ca] + K_{-1}/K_{+1}), where [Ca] is the concentration of Ca, and the time constant for the developing and declining K-creep would be given by ([Ca]$* K_{+1} + K_{-1}$)$^{-1}$ and $(K_{-1})^{-1}$, respectively.

Figure 3 shows a family of simulated K-creeps with several different K_{-1} values. Since K-creep at rest varies with K_{-1} (see below), ordinate denotes an extra creep conductance (gK-creep) recruited by the inward calcium current; the assumption was made that [Ca] amounts to 2.1 μM at time zero (0.1 μM in control: see Tatsumi et al. 1988). Hence, gK-creep can be expressed as a function of time (t) in an equation of the form

$$gK\text{-creep} = 21.5*(\exp(-t/\tau_{off}) - \exp(-t/\tau_{on}))$$

where τ_{on} and τ_{off}, respectively, denote ([Ca] $*K_{+1} + K_{-1})^{-1}$ and $(K_{-1})^{-1}$ while 21.5 (in nS) denotes the maximum gK-creep at time zero (North and Tokimasa 1987). K_{+1} =10^{6} M^{-1} s^{-1} is based on an assumption that it may be analogous to an association rate constant between ACh and nicotinic receptors. Absolute value for K-creep

North RA, Tokimasa T (1983) Depression of calcium-dependent potassium conductance of guinea-pig myenteric neurons by muscarinic agonists. J Physiol (Lond) 342:253-266

North RA, Tokimasa T (1984) The time course of muscarinic depolarization of guinea-pig myenteric neurones. Br J Pharmacol 82:085-091

North RA, Tokimasa T (1987) Persistent calcium-sensitive potassium current and the resting properties of guinea-pig myenteric neurones. J Physiol (Lond) 386:333-353

Pfaffinger P (1988) Muscarine and t-LHRH suppress M-current by activating an IAP-insensitive G-protein. J Neurosci 8:3343-3353

Smith PA (1994) Amphibian sympathetic ganglia: an owner's and operator's manual. Prog Neurobiol 43:439-510

Tatsumi H, Hirai K, Katayama Y (1988) Measurement of the intracellular calcium concentration in guinea-pig myenteric neurons by using fura-2. Brain Res 451:371-375

Tokimasa T (1985) Intracellular Ca^{2+}-ions inactivate K^+-current in bullfrog sympathetic neurons. Brain Res 337:386-391

Tokimasa T (1996) Calcium-dependent after-hyperpolarization in dissociated bullfrog sympathetic neurons. Neurosci Lett 218:49-52

Tokimasa T, Akasu T (1995) Biochemical gating for voltage-gated channels: mechanisms for slow synaptic potentials in autonomic ganglia. In: McLachlan EM (ed) Autonomic ganglia. Harwood,Readmg, pp 259-295

Tokimasa T, Nishimura T (1997) Actions of barium on rapidly inactivating potassium current in bullfrog sympathetic neurons. Neurosci Lett 236:37-40

Tokimasa T, North RA (1984) Calcium entry through acetylcholine-channel can activate potassium conductance in bullfrog sympathetic neurons. Brain Res 295:364-367

Tokimasa T, Cherubini E, North RA (1983) Nicotinic depolarization activates calcium dependent gK in myenteric Neurons. Brain Res 263:57-62

Tokimasa T, Morita K, North RA (1981) Opiates and clonidine prolong calcium-dependent after-hyperpolarizations. Nature 294:162-163

Tokimasa T, Shirasaki T, Kuba K (1997) Evidence for the calcium-dependent potentiation of M-current obtained by the ratiometric measurement of the fura-2 fluorescence in bullfrog sympathetic neurons. Neurosci Lett 236:123-126

Tokimasa T, Tsurusaki M, Akasu T (1991). Slowly inactivating potassium current in cultured bull-frog primary afferent and sympathetic neurones. J Physiol (Lond) 435:585-604

Tokimasa T, Ito M, Simmons MA, et al (1995) Inhibition by wortmannin of M-current in bullfrog sympathetic neurones Br J Pharmacol 114:489-495

Tokimasa T, Shirasaki T, Yoshida M, et al (1996a) Calcium-dependent potentiation of M-current in bullfrog sympathetic neurons. Neurosci Lett 214:79-82

Tokimasa T, Simmons M, Schneider CR, et al (1996b) Hyperpolarizing shift of the M-current activation curve after washout of muscarine in bullfrog sympathetic neurons. Neurosci Lett 207:97-100

Trouslard J, Marsh SJ, Brown DA (1993) Calcium entry through nicotinic receptors and calcium channels in cultured rat superior cervical ganglion cells J Physiol (Lond) 468:53-71

Slow Synaptic Responses in Neuronal Tumor Cells: Dual Regulation of ADP-Ribosyl Cyclase and Inhibition of M-Current by Muscarinic Receptor Stimulation

H. HIGASHIDA, S. YOKOYAMA, M. HASHII, and M. TAKETO

Department of Biophysical Genetics, Molecular Medicine and Bioinformatics, Kanazawa University Graduate School of Medicine, 13-1 Takara-machi, Kanazawa 920-8640, Japan

Key words. NAD+, CADP-ribose, Phospholipase C, Ca^{2+}, NG108-15 cells

Summary. Muscarinic acetylcholine receptors (mAChRs) utilize the direct signaling pathway to ADP-ribosyl cyclase via G proteins within cell membranes to produce cyclic ADP-ribose (cADPR) from β-NAD+. This signal cascade is analogous to the previously established transduction pathways from mAChRs to adenylyl cyclase and phospholipase Cβ. Together with cytosolic Ca^{2+}, cADPR functions to release Ca^{2+} through ryanodine receptors. This cADPR-dependent and mAChR-controlled increase in cytosolic Ca^{2+} concentrations may induce various cellular responses.

In neuroblastoma x glioma hybrid NG108-15 cells, it has been reported that bradykinin leads to cell membrane hyperpolarization followed by prolonged depolarization, during which action potentials are significantly inhibited and increased, respectively (Yano et al. 1984). The same response has been obtained by acetylcholine (ACh) in NG108-15 cells overexpressing m1- and m3-subtype of muscarinic ACh receptors (mAChRs; Fukuda et al. 1988). Voltage-clamp recordings reveal that the bradykinin- and ACh-induced hyperpolarization results from the activation of a Ca^{2+}-dependent K+ current ($I_{K(Ca)}$) via inositol-1,4,5-trisphosphate- (Ins(1,4,5)P$_3$)-induced Ca^{2+}, and the subsequent depolarization results from the inhibition of a voltage-dependent M-type K+ current (I_M), which is sensitive to muscarinic agonists (Higashida and Brown 1986; Ogura et al. 1990). Since protein kinase C activators mimic this I_M inhibition, we have proposed that I_M inhibition is due to protein kinase C-dependent phosphorylation of M-channels, as a result of diacylglycerol formation by phospholipase C activation by bradykinin or ACh (Higashida and Brown 1986; Brown and Higashida 1988). Our proposal of involvement of protein kinase C in I_M inhibition was criticized by Bomsa and Hille (1989), in that protein kinase C inhibitors can antagonize phorbol dibutyrate-induced I_M inhibition but not the agonist-induced inhibition in bullfrog ganglion cells. However, we did not test C-kinase inhibitors in NG108-15 cells,

Acetylcholine-Like Effect of Sulfhydryl-Modifying Reagents on M-Current in Rodent NG108-15 Cells

A.B. EGOROVA, N. HOSHI, and H. HIGASHIDA

Department of Biophysical Genetics, Molecular Medicine and Bioinformatics Kanazawa University Graduate School of Medicine, Takara-machi, 13-1, Kanazawa 920-8640, Japan

Key words. Muscarinic receptors, K^+ channels, GSH

Growing evidence has indicated that redox status regulates various aspects of cellular function through the reversible modification of redox-sensitive proteins in the signal cascade (Nakamura et al. 1997). The physiological significance of glutathione in the central nervous system (CNS) is still uncertain, although several data suggest a possible role for it as a neuromodulator/neurotransmitter, the existence of protein kinase C-regulated glutathione binding sites within the mammalian CNS (Lanius et al. 1994), and the depolarization-induced, Ca^{2+}-dependent release of glutathione and cysteine from rat brain slices have been reported (Zangerle et al. 1992). Modulation of ion channels by the cellular redox potential has emerged only recently, and it may provide a link whereby changes in the metabolic properties of a neuron can lead to changes in its electrical characteristics. The present study was undertaken to examine whether glutathione and different endogenous and exogenous sulfhydryl compounds modulate excitability in cholinergic neurons, which are known to be highly susceptible to oxidative stress and glutathione depletion (Li et al. 1998). To do this, effects of such reagents on voltage-dependent M channels were examined by measuring a voltage-dependent M-type potassium current ($I_{K(M)}$) in muscarinic acetylcholine receptor DNA (mAChR)-transfected neuroblastoma x glioma hybrid NG108-15 cells (Fukuda et al. 1988).

We found that external applications of glutathione in both reduced (GSH) and oxidized (GSSG) forms at millimolar concentrations in the acidic conditions induced a strong and reversible inhibition of $I_{K(M)}$ (79.5±5.1% and 75.4±9.3% of the control amplitude of M-current for 20 mM GSH and GSSG, respectively), giving rise to inward current in NG 108-15 cells (duration was from 10 s to 2 min, and the average amplitude was -0.72±0.12 nA and -1.24±0.22 nA for 20 mM GSH and GSSG, respectively). This effect was mimicked by external application of mercury and silver ions at a very low concentration of N-methylmaleimide

(NMM) and *p*-chloromercurybenzoate (pCMB). Dithiothreitol and 2-mercaptoethanol showed no direct effect on $I_{K(M)}$ by themselves and no blocking activity on the glutathione-induced response, suggesting that $I_{K(M)}$ inhibition results from modification of a protein thiol rather than of a disulfide bond (Egorova et al. 1997). Since thiol oxidation in our experiments was not due to the enzymatic reaction, the high concentration of glutathione required for reducing $I_{K(M)}$ may not be unreasonable. The shift in extracellular pH to the acidic level is the event involved in the pathogenesis of some neuronal disorders caused by action of reactive oxygen intermediates or found after increased excitability of neurones, thus providing the maximum reactivity for endogenous thiol compounds being released from the cell under conditions of oxidative stress or depolarization. The local change in the ratio of GSH and GSSG outside the cell, which was shown to be well correlated with the change in redox status of intracellular pyridine nucleotides (Weir et al. 1995), may alter the electrophysiological properties of potassium channels thereby adjusting their activity in accordance with the present metabolic conditions of the cell.

Table 1. Kinetic parameters of $I_{K(M)}$ in m1- subtype muscarinic AChR-transfected NG 108-15 cells treated with agents with M-current inhibitory activity

Treatment	Relative G_{max}, Δ	$V_{1/2}$(mV), Δ	k, Δ	τ_{deact} (ms), Δ
ACh,10 μM	0.40*	1.4	0.7	16
HgCl$_2$, 10 μM	0.28*	7.1	2.3	17
AgNO$_3$,10 μM	0.21*	6.1	2.2	5
pCMB,10 μM	0.27*	5.5	0.3	14
BaCl$_2$,1 mM	0.38*	14.5**	7.2**	48.1**

Relative permeability curves were fitted with the Boltzmann equation:
$$g = g_{max}/(1+\exp (V_{1/2} - V_c) / k)$$
to obtain values for the relative (g_{max}), the midpoint potential ($V_{1/2}$), and the slope parameter (k). For each parameter, values are the difference from the control (Δ), with n determinations ($n > 3$). Control values are 0.99 ± 0.02 for G_{max}, -60.4 ± 3.1 for $V_{1/2}$, 9.7 ± 0.95 for k, and 121 ± 7 for τ_{deact}. Steady-state currents were converted to conductances using the measured reversal potential and normalized to the control conductance at -10 mV. For more detail, see Table 2 (Egorova et al. 1997). τ_{deact} is the time constant obtained from the decay of tail currents at a test potential of -50 mV. *, significantly different from the control at $P < 0.005$; **, significantly different from the control at $P < 0.05$.

An interesting feature of $I_{K(M)}$ inhibition kinetics found in our study was that mercury, silver ions, pCMB, and acetylcholine (ACh) all behaved similarly, in a different manner from barium, the well known blocker of an M channel pore (Table 1), allowing us to speculate that the mechanism for inhibition of M-current by thiol-modifying agents may be the same as that by ACh. Sylfhydryl reagents may bind at an external site of M channels ouside the membrane electrical field

Inhibition of M-Type K⁺ currents by Cognition Enhancers in NG108-15 Cells and Rat Cerebral Neurons in Culture

M. NODA[1], H. HIGASHIDA[2], and N. AKAIKE[1]

[1]Department of Physiology, Kyushu University Faculty of Medicine, Fukuoka 812-8582
[2]Department of Biophysical Genetics Kanazawa University Graduate School of Medicine, Kanazawa 920-8640, Japan

Key words. AchR, M-current, KST-5452, Linopirdine, MG108-15 cells

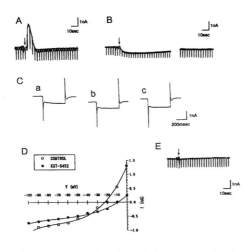

Fig. 1A-E. Effects of ACh and KST-5452 on potassium currents in control or m1-transformed NG108-15 cells. **A, B** Responses to focal application of ACh (**A**) and KST-5452 (**B**) in the same NG108-PM1-27 cell clamped at -20 mV; 3 ml of drug solution (1 mM concentration in the pipette) was applied at the *arrows*. Repetitive inward transients were evoked by step pulses to -40 mV for 0.4s. **C**: Expanded records of current transients obtained by -30 mV, 0.4s voltage steps before KST-5452 (*a*), during KST-5452-induced inward current (*b*), and after recovery (*c*) in a second NGPM1-27 cell. **D** Plots of the current attained at the ends of the voltage step as a function of the command voltage, before (*open circles*) and during (*filled circles*) the KST-5452-induced inward current. This steady-state current-voltage relationship was obtained in a third NGPM1-27 cell under identical recording conditions in **B**. **E**: Response to KST-5452 (3 ml, 1 mM) in a nontransfected NG108-15 cell clamped at -20 mV. Voltage command pulses, -30 mV for 0.4s.

The electrophysiological effects of KST-5452, an m1 muscarinic acetylcholine receptor (AChR) binding compound (Kishida et al. 1994), and linopirdine, a neurotransmitter release enhancer (Aiken et al. 1996; Nickolson et al. 1990) were studied in NG108-15 neuroblastoma x glioma hybrid cells transfected with m1-muscarinic AChR cDNA, using either the conventional whole-cell or the nystatin perforated-patch recording mode under voltage-clamp conditions.

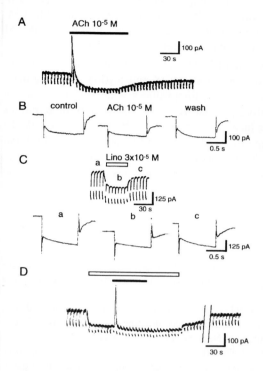

Fig. 2A-D. The inhibition of the M-current in NGPM1-27 cells by ACh and linopirdine. **A** ACh (10^{-5}M) induced a transient outward current ($I_{K(Ca)}$) followed by an inward current due to the inhibition of the M-current. **B** The inward current relaxation produced by the hyperpolarization from -20 to -40 mV for 1 s before (control), during (ACh 10^{-5} M), and after the application of ACh is shown. **C** Linopirdine (3 x 10^{-5} M) induced only the inward current due to the inhibition of the M-current. The holding current level returned to the original level after 5 min of washing out the linopirdine. The inward current relaxation produced by the hyper-polarization from -20 to -40 mV before, during, and after the application of linopirdine are shown in the *lower panel.* **D** In the presence of linopirdine, ACh still activated $I_{K(Ca)}$. An interval of ~60 s separates the trace into two parts. The holding potential was -20 mV and a hyperpolarizing 20 mV pulse of 1 s duration was applied every 5 s. ACh and linopirdine were applied during the period as indicated by the *horizontal bar* (the same as in **A** and **C**) above the current recording. **A, B, C,** and **D** were obtained from different cells.

Application of KST-5452 and linopirdine to m1-transformed NGPM1-27 cells elicited an inward current associated with decreased conductance and reduced M-type K^{+} current ($I_{K(M)}$) relaxation at a holding potential of -20 mV, which mimics an excitatory part of the ACh-induced responses in NGPM1-27 cells. The KST-5452-induced responses were blocked by pirenzepine, suggesting the KST-5452 acts as a potent excitant via m1-muscarinic AChRs at the concentration of 0.1-1 mM (Kishida et al. 1994). On the other hand, the effects of linopirdine were not blocked by atropine. The affinity of linopirdine for the inhibition of $I_{K(M)}$ was 24.7 μM in NGPM1-27 cells and 35.8 μM in pyramidal neurons acutely dissociated from the rat cerebral cortex. These values were slightly higher than that in the rat hippocampal pyramidal neurons and sympathetic ganglia (Aiken et al. 1996; Lamas et al. 1997), which has been reported to range from 3 to 8 μM. In the presence of linopirdine, ACh failed to evoke a further inward current, but ACh still elicited an outward current, thus suggesting that the Ca^{2+}-dependent K^+ current is rather insensitive to linopirdine (Noda et al. 1998). Our experiments also showed that linopirdine can inhibit other voltage-dependent K^+ currents ($I_{K(V)}$) in NGPM1-27 cells as well, with the concentration range less than 80 µM. Linopirdine is thus considered to be a relatively preferential inhibitor of the M channels, with a spectrum to other K channels. Interestingly, it has been reported that linopirdine inhibits the GABA-gated Cl^- currents and nicotinic acetylcholine-gated ionic channels at similar low concentrations (8 and 5 µM) in sympathetic ganglia (Lamas et al. 1997), thus indicating linopirdine's wide spectrum for ion channels. In conclusion, these results show that KST-5452 and linopirdine mimics the excitatory action of the ACh-induced responses in m1-transformed NG108-15 cells. Also, the cognition enhancing effects of KST-5452 (Kishida et al. 1994) and linopirdine are mediated through an inhibition of $I_{K(M)}$, and thereby the physiological importance of $I_{K(M)}$ in the central nervous system would be indicated.

References

Aiken SP, Zaczek R, and Brown BS (1996) Pharmacology of the neurotransmitter release enhancer linopirdine (DuP 996), and insights into its mechanism of action. Adv Pharmacol 35:349-384

Kishida H, Yamamoto K, Fuse Y, et al (1994) Activation of inward current associated with M-potassium current inhibition in m1-muscarinic receptor-transformed NG108-15 cells by KST-5452, a novel cognition enhancer. Neurosci Lett 172:119-121

Lamas JA, Selyanko AA, Brown DA (1997) Effects of a cognition-enhancer, linopirdine (DuP 996), on M-type potassium currents $I_{(K(M))}$ and some other voltage- and ligand-gated membrane currents in rat sympathetic neurons. Eur J Neurosci 9:605-616

Nickolson VJ, Tam SW, Meyers MJ, et al (1990) DuP 996 (3, 3-bis(4-pyridinylmethyl)-1-phenylindolin-2-one) enhances the stimulus-induced release of acetylcholine from rat brain in vitro and in vivo. Drug Dev Res 19:285-300

Noda M, Obana M, Akaike N (1998) Inhibition of M-type K^+ current by linopirdine, a neurotransmitter release enhancer, in NG108-15 neuronal cells and rat cerebral neurons in culture. Brain Res 794:274-280

Muscarinic Inhibition of M-current in Bullfrog Sympathetic Neurones is Independent of Intracellular Ca^{2+} Release

T. AKITA and K. KUBA

Department of Physiology, Nagoya University School of Medicine, 65 Tsurumai-cho, Showa-ku, Nagoya, 466-8550, Japan

Key words. Muscarinic receptor, M-current, Cyclic ADP-ribose, Streptozocin, Ca^{2+} release and Bullfrog sympathetic ganglion cells.

Summary. Possible roles of cyclic ADP ribose (cADPR) and intracellular Ca^{2+} release in the action of muscarine to inhibit a voltage-dependent, noninactivating K^{+} current, I$_M$, were studied in bullfrog sympathetic ganglion cells. cADPR applied intracellularly via a patch pipette affected neither I$_M$ nor the inhibitory action of muscarine on I$_M$. Muscarine produced a rise in [Ca^{2+}]$_i$ only in 1 cell of 5, but consistently suppressed I$_M$. Neither ryanodine nor thapsigargin affected the inhibitory action of muscarine on I$_M$ and muscarine-induced rise in [Ca^{2+}]$_i$. Thus, intracellular Ca^{2+} release and cADPR are not involved in the signal transduction from muscarinic receptors I$_M$ in bullfrog sympathetic ganglion cells.

1 Introduction

Activation of muscarinic receptors suppresses a voltage-dependent, non-inactivating K current (I$_M$) and generates slow excitatory postsynaptic potentials in many types of neurones. The mechanism linking muscarinic receptors and the K^{+} channel involved in I$_M$ (M-channel), however, is still unknown. One possible mechanism suggested by a study on neuroblastoma-glioma cells is that cyclic ADP-ribose (cADPR) produced from βNAD mediates the information from muscarinic receptors to M-channels (Higashida et al. 1995). cADPR is known to modulate Ca^{2+} release from intracellular Ca^{2+} stores (Galione et al. 1991; Hua et al. 1994). We examined the role of cADPR and [Ca^{2+}]$_i$ in modulation of I$_M$ in cultured bullfrog sympathetic ganglion cells.

2 Methods

We used the conventional whole-cell patch-clamp technique to record I_M. The composition of Ringer's solution was as the following (in mM): NaCl, 115; KCl, 2.5; $MgCl_2$, 1; $CaCl_2$, 2; HEPES, 5; glucose, 10 (pH 7.2). Patch electrode solution contained (in mM): K-gluconate or aspartate, 130; KCl, 7; HEPES, 10; $MgCl_2$, 2; Na_2ATP, 4; Na_3GTP, 0.3; EGTA, 1 (pH 7.35). For $[Ca^{2+}]_i$ measurement, the cell was loaded with 10 µM Oregon Green BAPTA-1 K via a patch electrode and then its fluorescence intensity was monitored with a laser-scannning confocal imaging system (Bio-Rad MRC-600). For that purpose, EGTA was excluded from the patch electrode solution.

3 Results and Discussion

To investigate the effect of cADPR on I_M inhibition, we applied 2 µM cADPR into cultured cells via patch pipettes. As opposed to Higashida et al (1995), the rate of I_M rundown during the course of application of cADPR (22 ± 8.5 at 30 min, mean% ± SEM, $n = 6$) was similar to that of control (38 ± 6.6, $n = 6$). The degree of I_M suppression by muscarine (74.75 ± 8.02, $n = 4$) was not significantly different from that of control (87.7 ± 3.47, $n = 6$), either. Raising the concentration of cADPR to 100 µM was not effective to enhance the effect of muscarine ($n = 1$). Preincubation of the cells with 20 µM streptozotocin for 6-12 hrs, known to reduce the level of βNAD (a precursor of cADPR), had no effect on I_M and the magnitude of the muscarinic action on I_M (78 ± 3.12, $n = 14$). These results suggest that cADPR is not responsible for the muscarinic action of I_M as a modulator in bullfrog sympathetic neurons. Surprisingly, a certain fraction of cells showed augmentation of I_M by 10 µM muscarine (120.25 ± 3.52, $n = 12$). cADPR, however, did not change the degree of this augmentation (115.67 ± 5.57, $n = 6$).

We also examined the role of intracellular Ca^{2+} release in the muscarinic action of I_M. Muscarine (10 µM) evoked a transient, homogeneous rise in $[Ca^{2+}]_i$ in some cells ($n = 1$), but not in other cells ($n = 4$), when the cell membrane was held at − 30 mV. The inhibition of I_M by muscarine, however, was consistently observed and maintained as long as the drug was present. Ryanodine (10 µM) blocked the transient rise in $[Ca^{2+}]_i$, but did not affect the muscarinic inhibition of I_M ($n = 1$). The muscarine-induced depolarization was not accompanied by a rise in $[Ca^{2+}]_i$ in our preparations ($n = 2$). Thapsigargin(1µM) did not affect the muscarine-induced depolarization of the cells and did not alter the $[Ca^{2+}]_i$ in the course of muscarine application ($n = 1$).

Our study indicates that neither intracellular Ca^{2+} release nor the action of cADPR be involved in the muscarinic suppression of I_M in bullfrog sympathetic ganglion cells.

References

Galione A, Lee HC, Busa WB (1991) Ca^{2+}-induced Ca^{2+} release in sea urchin egg homogenates: Modulation by cyclic ADP-ribose. Science 253:1143-1146

Higashida H, Robbins J, Egorova A, et al. (1995) Nicotinamide-adenine dinucleotide regulates muscarinic receptor-coupled K^+ (M) channels in rodent NG 108-15 cells. J Physiol 482.2:317-323

Hua SY, Tokimasa T, Takasawa S, et al. (1994) Cyclic ADP-ribose modulates Ca^{2+} release channels for activation by physiological Ca^{2+} entry in bullfrog sympathetic neurons. Neuron 12:1073-1079

oocyte expressing IRK3 as well as m1 AChR, which was stimulated by carbachol, indicating that the m1 receptor-mediated channel inhibition is superimposed on the tonic inhibition of IRK3 channels. Unlike IRK3, the IRK1 current showed little increase upon excision of the membrane patch from the oocyte, indicating that IRK1 channels were scarcely inhibited by cytoplasmic messengers inside the oocyte.

Given that IRK3 but not IRK1 channels are sensitive to m1 modulation as well as tonic inhibition by a cytoplasmic factor, we wondered whether the same factor mediating tonic inhibition is also responsible for channel inhibition by the m1 receptor. In this scenario, the inhibitory factor is present in the unstimulated oocyte and causes tonic inhibition; m1 receptor stimulation simply raises the concentration of the inhibitory factor and causes further suppression of current.

Tonic inhibition of IRK3 was observed not only in *Xenopus* oocytes but also in mammalian expression systems such as CHO (Chinese hamster ovary) cells and HEK (human embryonic kidney) 293 cells transfected with IRK3 (Fig. 3). The IRK3 current recorded in this whole-cell configuration gradually increased with time, revealing channel inhibition by a cytoplasmic factor that may be slowly lost into the pipette solution. Inclusion of protein phosphatase inhibitors such as okadaic acid, orthovanadate, or the tyrosine kinase inhibitor genestein both in the bath and in the pipette solution did not prevent the IRK3 current from increasing during whole-cell recording. This indicates that tonic inhibition was not due to channel phosphorylation. A similar conclusion was also obtained for tonic inhibition in the oocyte, when we applied these phosphatase inhibitors to IRK3 channels in inside-out membrane patches. Inclusion of ATP, AMPPNP, AMPPCP, or ADP in the pipette solution also did not prevent the IRK3 current from increasing during whole-cell recording, indicating that the tonic inhibition could not be caused by channel inhibition by ATP.

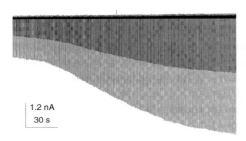

Fig. 3. Continuous whole-cell recording of a mammalian cell expressing IRK3 showed the enhancement of current by including EDTA in the pipette solution.

1.2 nA
30 s

Whether the IRK3 current increased in amplitude during whole-cell recording depended on the type of chelators present in the pipette solution (Fig. 3). The increase of IRK3 current was observed when the cytosol came in contact with pipette solution containing either EDTA or CDTA, but not when the cytosol was exposed to pipette solution containing EGTA. The major difference between EDTA and EGTA is the ability of the former but not the latter to chelate Mg^{2+}

This led us to investigate whether Mg^{2+} exerts any inhibitory effect on IRK3 channels as well as the chimeras and IRK1 mutants that are sensitive to m1 modulation.

Mg^{2+} applied to inside-out patches from *Xenopus* oocytes had a concentration-dependent inhibitory effect on IRK3. By contrast, IRK1 was essentially not inhibited up to 2 mM Mg^{2+}. Because the m1 modulation of IRK3 current was also accompanied with an enhancement of inward rectification, we wanted to determine if the Mg^{2+} inhibition of IRK3 is voltage dependent. Mg^{2+} inhibition of IRK3 involved both inward and outward current, had a slow time course, and was independent of membrane potential. This inhibitory effect of Mg^{2+} is therefore distinguishable from the rapid and voltage-dependent action of Mg^{2+} as a pore blocker.

The increase of inward rectification of IRK3 upon m1 stimulation urged us to look at some IRK1 rectification mutants (Yang et al. 1995). Among the three IRK1 mutants that have lower affinities for Mg^{2+} and polyamines, one of them became sensitive to m1 modulation. This double mutant, IRK1 D172NE224G, has its mutations outside the domains we mapped from the chimera study. When we tested the Mg^{2+} sensitivity of this mutant, the dose-response curve was as steep as IRK3 but the EC_{50} was increased. Also, we only saw marginal tonic inhibition with this channel. In contrast, the minimal chimera that endows channels sensitivity to m1 modulation was inhibited by lower concentration of Mg^{2+}, but with a less steep dose-response curve.

The inhibitory effect of Mg^{2+} on IRK3 and the mutant channels argues for the involvement of a common structure in these channels for both Mg^{2+} inhibition and m1 modulation, and suggests that Mg^{2+} is responsible for both of them. It can also account for the discrepancy between the level of tonic inhibition (minimal chimera > IRK3 > IRK1 D172NE224G) and the sensitivity of these channels to m1 modulation (minimal chimera < IRK1 D172NE224G < IRK3). However, we could not measure a rise of Mg^{2+} concentration after m1 receptor stimulation (a 200-300 μM rise predicted from the dose-response curves). It remains possible that an as yet unidentified messenger that can behave in a similar fashion as Mg^{2+} is the authentic mediator of m1 modulation or m1 receptor activation can lead to some channel modification that changes the sensitivity of these channels to the inhibitory action of Mg^{2+}. Recently it has been shown that phosphatidylinositol bisphosphate (PIP_2) can stimulate several inward rectifier K^+ channels in the excised membrane patches (Fan and Makielski 1997; Hilgemann and Ball 1996; Huang et al. 1998). Because one of the effectors of m1 receptor stimulation is phospholipase Cβ, it will be interesting to see if the reduction of PIP_2 level will change the sensitivity of channels to Mg^{2+} and how PIP_2 will be involved in the tonic inhibition of the IRK channels.

Having found that IRK3 channels are subject to tonic inhibition by Mg^{2+} inside the cell, we did single channel recording to determine the mechanism that contributes to this tonic inhibition. We found that in the cell-attached configuration the IRK3 channels in the membrane patch tend to enter a prolonged inactivated state (lasting for 1-5 minutes for channels that we observed subsequent openings). Usually two to three of these long inactivation events took place within 10 min of continuous recording. Occasionally we observed that all the channels in a patch entered the long inactivated state, resulting in brief quiescence without any channel activity. Figure 4 is an example of such a patch with at least five channels, although for most of the time there were at most two channels active simultaneously. After the membrane patch was excised from the cell, the IRK3 channels no longer entered the inactivated state readily, leading to almost full activation of every channel within the patch. This is illustrated in the histogram in Figure 4 for recordings from an oocyte membrane patch containing at least four channels.

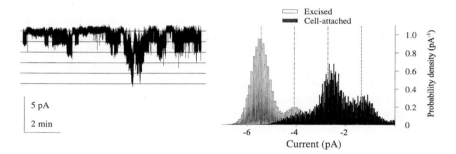

Fig. 4.   Left panel: A long inactivation state of IRK3 causes tonic inhibition in the cell-attached configuration. In a patch with at least five channels, usually fewer than two channels were active simultaneously. Right panel: Comparison of the amplitude histograms between the cell-attached and the excised inside-out configurations (into EDTA) of a patch revealed the tonic inhibition of IRK3 by cytoplasmic Mg^{2+}.

Like IRK3, the minimal chimera also entered a long inactivated state quite frequently. In contrast, IRK1 rarely entered a long inactivated state, the longest duration we have measured for an IRK1 channel to enter the inactivated state was less than 30 s. Thus, IRK3 but not IRK1 channels are susceptible to m1 modulation and tonic inhibition, which could be accounted for by the ability of Mg^{2+} to promote the long-lasting inactivation.

Mg^{2+} is the third most abundant cation inside the cell. Most of the Mg^{2+} ions inside the cell are either sequestered in internal organelles or bound by cytosolic proteins. The rest is mostly complexed with small organic molecules such as nucleotides and metabolic intermediates including citric acid. Consequently, free Mg^{2+} represents only 2-3% of the total cellular Mg^{2+}. Epinephrine and phorbol ester have been found to regulate Mg^{2+} influx and hence alter free cytoplasmic Mg^{2+} level (Elliott and Rizack 1974; Erdos and Maguire 1983; Grubbs et al. 1984; Grubbs and Maguire 1986). However, since m1 AChR-mediated inhibition of IRK3 channel activity persisted in the absence of external Mg^{2+}, it seems unlikely that the m1 AChR alters intracellular Mg^{2+} concentration by regulating Mg^{2+} influx, thereby modulating IRK3 channel activity. Therefore, the free Mg^{2+} concentration change sensed by the channels must be from at least one of the following sources: the intracellular organelles that store Mg^{2+}, the cytoplasmic buffers or binding proteins that sequester Mg^{2+}, or the functional compartments that hinder free diffusion of Mg^{2+}.

A small and slow change in free Mg^{2+} is better suited for setting the gain of a system, which is usually achieved through continual adjustment of the system to its various inputs over a relatively long period. For Mg^{2+} to serve as a messenger, the response element has to be able to sense a small change in Mg^{2+} concentration. The IRK3 channels actually meet this requirement by showing a remarkable sensitivity to free Mg^{2+} concentration change.

The m1 AChR-mediated inhibition of IRK3 channel activities persists for a few minutes after application and subsequent removal of agonists and will be suitable for the fine tuning of membrane resistance on a minute-to-minute or even slower basis. For tonic neurons in the sympathetic ganglia and many central neurons, fast synaptic inputs produce only small, subthreshold EPSPs that have to summate to fire an action potential (Crowcroft et al. 1971; McLachlan and Meckler 1989). Integration of synaptic inputs may be more effectively regulated by a slow but persistent inhibition of inward rectifier K$^+$ channels, as observed in this study.

References

Chuang H, Jan YN, Jan LY (1997) Regulation of IRK3 inward rectifier K$^+$ channel by m1 acetylcholine receptor and intracellular magnesium. Cell 89:1121-1132

Crowcroft PJ, Holman ME, Szurszewski JH (1971) Excitatory input from the distal colon to the inferior mesenteric ganglion in the guinea pig. J Physiol (Lond) 219:443-461

Elliott DA, Rizack MA (1974) Epinephrine and adrenocorticotropic hormone-stimulated magnesium accumulation in adipocytes and their plasma membranes. J. Biol. Chem. 249:3985-3900

Erdos JJ, Maguire ME (1983) Hormone-sensitive magnesium transport in murine S49 lymphoma cells: characterization and specificity for magnesium. J. Physiol. 337:351-371

Fakler B, Brandle U, Glowatzki E, et al (1995) Strong voltage-dependent inward rectification of inward rectifier K$^+$ channels is caused by intracellular spermine. Cell 80:149-154

Fan Z, Makielski JC (1997) Anionic phospholipids activate ATP-sensitive potassium channels. J Biol Chem 272:5388-5395

Ficker E, Taglialatela M, Wible BA, et al (1994) Spermine and spermidine as gating molecules for inward rectifier K$^+$ channels. Science 266:1068-1072

Grubbs RD, Collins SD, Maguire ME (1984) Differential compartmentation of magnesium and calcium in murine S49 lymphoma cells. J. Biol. Chem. 259:12184-12192

Grubbs RD, Maguire ME (1986) Regulation of magnesium but not calcium transport by phorbol ester. J. Biol. Chem. 261:12550-12554

Hilgemann DW, Ball R (1996) Regulation of cardiac Na^+,Ca^{2+} exchange and K_{ATP} potassium channels by PIP_2. Science 273:956-959

Huang CL, Feng S, Hilgemann DW (1998) Direct activation of inward rectifier potassium channels by PIP_2 and its stabilization by $G\beta\gamma$. Nature 391:803-806

Lopatin AN, Makhina EN, Nichols CG (1994) Potassium channel block by cytoplasmic polyamines as the mechanism of intrinsic rectification. Nature 372:366-369

Matsuda H (1991) Magnesium gating of the inwardly rectifying K^+ channel. Annu Rev. Physiol. 53:289-298

Matsuda H, Saigusa A, Irisawa H (1987) Ohmic conductance through the inwardly rectifying K channel and blocking by internal magnesium. Nature 325:156-159

McLachlan EM, Meckler RL (1989) Characteristic of the synaptic input to three classes of sympathetic neurone in the coeliac ganglion of the guinea-pig. J Physiol (Lond) 415:109-129

North RA, Uchimura N (1989) 5-Hydroxytryptamine acts at $5\text{-}HT_2$ receptors to decrease potassium conductance in rat nucleus accumbens neurones. J Physiol (Lond) 417:1-12

Stanfield PR, Davies NW, Shelton PA, et al (1994) The intrinsic gating of inward rectifier K^+ channels expressed from the murine IRK1 gene depends on voltage, K^+ and Mg^{2+}. J Physiol (Lond) 475:1-7

Takano K, Stanfield PR, Nakajima S, et al (1995) Protein kinase C-mediated inhibition of an inward rectifier potassium channel by substance P in nucleus basalis neurons. Neuron 14:999-1008

Uchimura N, Cherubini E, North RA (1989) Inward rectification in rat nucleus accumbens neurons. J Neurophysiol 62:1280-1286

Uchimura N, North RN (1990) Muscarine reduces inwardly rectifying potassium conductances in rat nucleus accumbens neurones. J Physiol (Lond) 422:369-380

Vandenberg CA (1987) Inward rectification of a potassium channel in cardiac ventricular cells depends on internal magnesium ions. Proc Natl Acad Sci USA 84:2560-2564

Wang H-S, McKinnon D (1996) Modulation of inwardly rectifying currents in rat sympathetic neurones by muscarinic receptors. J Physiol (Lond) 492:467-478

Yang J, Jan YN, Jan LY (1995) Control of rectification and permeation by residues in two distinct domains in an inward rectifier K^+ channel. Neuron 14:1047-1054

Temporal Profile of Muscarinic Modulation of the Slow Ca^{2+}-Dependent K^+ Current (I_{sAHP}) in Rat Hippocampal Neurons

L. ZHANG[1,2,5], A.A. VELUMIAN[1], P.S. PENNEFATHER[1,3,4], and P.L. CARLEN[1,2,3,5]

[1]Playfair Neuroscience Unit, Toronto Hospital Research Institute
Departments of [2]Medicine (Neurology) and [3]Physiology, [4]Faculty of Pharmacy, and
[5]Bloorview Epilepsy Program, University of Toronto, Toronto, Ontario, Canada

Key words. Acetylcholine, Acetylcholinesterase, Acetylcholinesterase inhibitor, Anions, Brain slice, Calcium, Calcium chelator, Calcium-dependent potassium channels, Hippocampus, Kinetics, Muscarinic, Synapse, Whole-cell Patch-clamp recording

1 Introduction

The central muscarinic cholinergic system plays a critical role in cognitive functions, and impairment of this system has been implicated in neurodegenerative disorders such as Alzheimer disease (Everitt and Robbins 1997; Whitehouse et al. 1981; Winkler et al. 1995; Wurtman 1992; Zola-Morgan and Squire 1993). Despite the accumulating evidence showing muscarinic modulation of synaptic and ionic activities of central neurons, we know little about the kinetics of this modulation, particularly in physiological conditions where muscarinic responses are elicited following the synaptic release of acetylcholine. Examination of the kinetics of muscarinic modulation is important for understanding possible dysfunction of muscarinic cholinergic synapses (Hsu et al. 1997; Taylor and Griffith 1993) and for evaluating the effects of acetylcholinesterase (AChE) inhibitors designed to treat patients with Alzheimer disease (Eagger and Harvey 1995; Parnetti 1995; Poirier et al 1995; Wagstaff and McTavish 1994). Moreover, because the commonalties exist in signal transduction pathway of metabotropic G protein-coupled receptors (Wess 1993), the kinetics of muscarinic cholinergic effects may have general implications for this class of neurotransmitter receptor-mediated systems.

The hippocampus, which plays a role in memory functions, receives cholinergic inputs originating from the medial septum and the diagonal band of

Broca (Lewis and Shute 1967) and expresses high levels of muscarinic receptors (Hulmel et al 1990; Levey et al. 1995). Stimulation of muscarinic receptors induces profound changes in synaptic activities and intrinsic ionic conductances of hippocampal neurons, particularly causing the inhibition of a Ca^{2+}-dependent K^+ current (I_{sAHP}) that underlies the slow afterhyperpolarization (sAHP) following repetitive discharges (Dutar et al. 1995; Krnjevic 1993; Storm 1990; see also the chapters by D.A. Brown, H.H. Chuang, H Tsubokawa, or J.F. Storm, this volume). The sAHP/I_{sAHP} is highly sensitive to muscarinic stimulation, with an EC_{50} value of 0.3 μM for the reduction of sAHP/I_{sAHP} by carbachol (Madison et al. 1987).

The reduction of the I_{sAHP} is not associated with decreases in corresponding intracellular Ca^{2+} signals or Ca^{2+} currents (Knöpfel et al. 1990; Müller and Connor 1991; Tsubokawa and Ross 1996; Zhang et al 1996), likely reflecting the modulation of the underlying K^+ channels by kinase-dependent processes (Abdul-Ghani et al. 1996; Baskys et al. 1990; Müller et al. 1992; Pedarzani and Storm 1993, 1996; Sah and Isaacson 1995). Moreover, the muscarinic reduction of the sAHP/I_{sAHP} in hippocampal neurons is readily induced in brain slices by electrical stimulation of cholinergic afferent fibers (Cole and Nicoll 1983, 1984; Krnjevic 1993; Zhang et al. 1995), allowing a direct assessment of functionality of muscarinic cholinergic synapses and the resulting downstream cascades.

In an attempt to reveal the kinetics of muscarinic modulations in central neurons, we have examined the temporal profile of the reduction of the I_{sAHP} following stimulation of cholinergic afferent fibres in rat brain slices. Our results demonstrated that in rat hippocampal CA1 neurons, an evoked release of ACh is capable of reducing the I_{sAHP}, providing that it is given within 2 s before the activation of the I_{sAHP} (Figure 1a). The cholinergic input is without effect on the following I_{sAHP} when it is given immediately after the onset of this current. We hypothesize that in physiological conditions, muscarinic modulation of central neurons may have a limited time window following the cholinergic impulse, and that the K^+ channels underlying the Ca^{2+}-dependent I_{sAHP} are inhibited by muscarinic receptor initiated signaling cascades only in the inactive state, i.e., before their activation by the elevated intracellular Ca^{2+}.

2 Effects of Internally Applied Anions or Calcium Chelators on the sAHP/I_{sAHP}

When considering the neurotransmitter modulation of the sAHP/I_{sAHP}, it is perhaps necessary to first discuss the issue of how to stabilize the Ca^{2+}-dependent sAHP/I_{sAHP} in the whole-cell patch-clamp recordings. The use of patch-clamp recording techniques in brain slices has greatly promoted electrophysiological studies in mammalian central neurons. The main drawback of this approach is the rundown of ionic conductances following extensive whole-cell dialysis (Bean 1992; Kay 1992), particularly of the Ca^{2+}-dependent ones including the sAHP/I_{sAHP}. To establish an intracellular milieu that supports the sAHP/I_{sAHP}

during the whole-cell dialysis, we first screened the effects of internally applied anions by using a patch pipette solution that had 2 mM Mg-ATP, 2 mM HEPES, 0.1 mM EGTA, and 150 mM of several potassium salts.

Effects of several potassium salts were examined including potassium chloride, potassium methylsulfate (KMeSO$_4$), potassium gluconate (KGluc), potassium citrate, potassium glutamate, potassium acetate, and potassium sulfate. The patch pipette solution made with KMeSO$_4$ was most supportive for whole-cell recordings of the sAHP/I$_{sAHP}$ (Zhang et al. 1994). In hippocampal neurons dialyzed with KMeSO$_4$ at 32-33 °C, I$_{sAHP}$s could be repeatedly evoked for more than 30 min, and their amplitude and time course were closely comparable to those previously recorded via using sharp electrodes. In contrast, hippocampal neurons showed either a rapid rundown or complete absence of the sAHP/I$_{sAHP}$ during whole-cell dialysis with other potassium salts (Zhang et al. 1994). A similar trend was also found in whole-cell study of hypoxia-induced changes in hippocampal neurons of rat brain slices, where a brief hypoxic episode induces a profound K$^+$ current in neurons dialysed with KMeSO$_4$, but not with KGlu or KCl (Belousov et al. 1996; Zhang and Krnjevic 1993; Chung et al. 1998).

To further reveal the interactions of internally applied anions and the Ca^{2+}-dependent sAHP/I$_{sAHP}$, we established an intracellular perfusion system that permits fast and repeated exchanges between the patch pipette solution and the cytoplasm of the recorded hippocampal neurons of rat brain slices (Velumian et al 1993). Interestingly, the deterioration of the sAHP/I$_{sAHP}$ following whole-cell dialysis with KGluc, one potassium salt that is widely used in whole-cell recording experiments, recovered when it was replaced with KMeSO$_4$ (Velumian et al. 1997). Moreover, the depressed sAHP/I$_{sAHP}$ recorded in the presence of intracellular KGluc was enhanced by intracellular perfusion of BAPTA, a fast and high affinity Ca^{2+} chelator (Velumian et al. 1997). The results obtained from intracellular perfusion studies suggest that the activation of the sAHP/I$_{sAHP}$ and the underlying intracellular Ca^{2+} signals is a labile process, and that the functional state of this process is strongly influenced by whole-cell recording conditions. The alteration induced by internally applied anions may generalize to other ionic conductances or to neurons of other brain regions (Velumian et al. 1997). Caution should be used in future studies to consider the possible changes in cell and signalling pathway physiology resulting from internally applied anions or calcium buffers during whole-cell dialysis.

3 The I$_{sAHP}$ Reduction by Afferent Stimulation is Mediated by Muscarinic Receptors

To examine the effects of cholinergic stimulation on the I$_{AHP}$, hippocampal CA1 neurons of adult rat brain slices were recorded in the whole-cell mode using the KMeSO$_4$-containing patch pipette solution. The neurons were voltage-clamped at potentials near -55 mV, and the I$_{AHP}$s were evoked by constant depolarizing pulses

(200-300 ms, 50-60 mV) every 30 s. Cholinergic afferent stimulation was achieved by placing a bipolar tungsten electrode in stratum oriens close to the whole-cell recording site (Cole and Nicoll 1983,1984), and the stimulation strength was adjusted to ensure a substantial decrease in the I_{sAHP} without spiking. Under these conditions, the afferent stimulus, when applied within 2 s before the depolarizing pulse, caused a profound decrease in the evoked I_{sAHP}. Inhibition of the I_{sAHP} reversed in 30-60 s, and it could be repeated many times in the same neurons. The I_{sAHP} reduction following the afferent stimulus persisted in the presence of 6-cyano-7-nitroquinoxaline-2,3-dione (CNQX, 20 µM), D-2-amino-5-phosphonopentanoic acid (D-AP, 50 µM) and 10 µM bicuculline methiodide, which block ionotropic glutamate and $GABA_A$ receptors respectively. The I_{sAHP} reduction following the afferent stimulus was fully abolished after perfusion of slices with 1-5 µM atropine.

These results are consistent with early studies (Cole and Nicoll 1983, 1984; Krnjevic 1993; Zhang et al. 1992), confirming a monosynaptic activation of muscarinic receptors that is responsible for inhibiting the I_{sAHP} following the afferent stimulation (Zhang et al. 1995,1996). Reduction of the Ca^{2+}-dependent I_{sAHP} following afferent stimulation is unlikely resulted from diminished Ca^{2+} entry, because neither the afferent stimulation nor external application of 5-10 µM carbachol suppressed depolarization-activated inward Ca^{2+} currents in hippocampal CA1 neurons (Zhang et al. 1996). Other studies using simultaneous intracellular recordings and fluorescent Ca^{2+} imaging techniques have shown that the reduction of the I_{sAHP} is not associated with decreases in the corresponding intracellular Ca^{2+} signals (Knöpfel et al. 1990; Müller and Connor 1991). Collectively, these results suggest that the modulation of the Ca^{2+}-dependent sAHP/I_{sAHP} may take place at sites downstream from depolarization-gated Ca^{2+} entry.

4 Time Window Permits the Reduction of the I_{sAHP} Following the Afferent Stimulation

Two paradigms were used to determine the time frame in which the I_{sAHP} is decreased following the afferent stimulus. First, we evoked the I_{sAHP}s repeatedly using four depolarizing pulses over a period of 10 s. The time intervals separating the first pulse and following pulses were 2, 5, and 10 s respectively. Once stable I_{sAHP}s were recorded such that the currents evoked by the first, third, or fourth pulses were comparable in their amplitudes, and cholinergic afferent fibers were stimulated electrically 200 ms before the first depolarizing pulse. Reduction in the I_{sAHP} was revealed by subtracting the current records obtained before and immediately after the afferent stimulus. Subtracted responses, which represented the net reduction of the I_{sAHP}, were largest following the first depolarizing pulse and barely detectable in response to the fourth pulse. In a set of four CA1 neurons examined, the normalized reduction in the I_{sAHP} was 43.0±2.8%, 21.5±3.2%, 4.8±3.1%, or 2.4±1.3%, respectively, for I_{sAHP}s evoked at 0.2, 2, 5, or 10 s following the afferent stimulus (Zhang et al. 1997).

Using this paradigm, the changes in the I_{sAHP} following the afferent stimulus can be examined sequentially in a short period. One concern about this paradigm is that the intracellular Ca^{2+} signal corresponding to the I_{sAHP} generation may not recover fully before the next signal is evoked, thereby interfering with the muscarinic cascades responsible for the I_{sAHP} reduction. To clarify this issue, we used another paradigm. The I_{sAHP}s were evoked by depolarizing pulses every 30 seconds to ensure the recovery of intracellular Ca^{2+} signals, and the afferent stimulus was applied 1-30 s before the depolarizing pulse. The I_{sAHP}s recorded before and after the afferent stimulus were compared in the same neuron. With this paradigm, the degree of I_{sAHP} reduction following the afferent stimulus dropped steeply with increasing time between the stimulus and the depolarizing pulse that triggered the I_{sAHP}. When a similar stimulus was given at 1, 2, or 5 s before to the depolarizing pulse, the following I_{sAHP} was decreased by 49.3±6.8%, 14.3±1.9%, or 6.8±1.9% (n=6), respectively, from the baseline control. The afferent stimulus caused no decrease in the subsequent I_{sAHP} when it was given ≥10 s before the depolarizing pulse. The relation between the I_{sAHP} reduction and the time separating the stimulus and the following depolarizing pulses could be described by a single exponential fit, with a time constant of 1.45 s. These results suggest a limited time window of ~2 s in which the cholinergic afferent stimulus could effectively decrease the evoked I_{sAHP} (Zhang et al. 1997)..

Acetylcholine (ACh) is a labile transmitter, and its ambient level in mammalian cerebrospinal fluid is in the low nM range (Flentge et al. 1992). The low level of ACh in the brain is due to the high rate of AChE hydrolysis of ACh (K_{cat}=1.6x10^4 s^{-1}; Massoulié et al. 1993; Taylor and Radic 1994). To determine the degree to which the efficacy of the cholinergic afferent stimulation is controlled by AChEs under our experimental conditions, we perfused brain slices with 9-amino-1,2,3,4-tetrahydro-acridine (THA) or physostigmine. These two agents are centrally acting inhibitors known to suppress the catalytic activity of AChEs (Eagger and Harvey 1995; Parnetti 1995; Poirier et al 1995; Taylor and Radic 1994; Wagstaff and McTavish 1994). In the presence of THA or physostigmine, the effective time frame in which the cholinergic afferent stimulation decreased the evoked I_{sAHP} was expanded ≥10fold. Such that cholinergic afferent stimulus decreased a subsequent I_{sAHP} by 18% (n=7) even when it was given 30 seconds before the depolarizing pulse that triggered the I_{sAHP}. The EC_{50} for THA to enhance the efficacy of the afferent stimulus was 0.4 μM (Zhang et al. 1997). These results are consistent with in vivo findings (see earlier references), providing direct evidence that the strength of central cholinergic synapses is controlled by the activity of AChEs.

5 Temporal Specificity of Muscarinic Reduction of the I_{sAHP}

Previous studies using simultaneous intracellular recordings and fluorescent Ca^{2+} imaging have shown that the long-lasting I_{sAHP} is paralleled by a rise of intracellular Ca^{2+} in the range of 50 to 200 nM. Interestingly, the transmitter-induced reduction of the I_{sAHP} is not associated with a decrease in the corresponding Ca^{2+} signals (Knöpfel et al. 1990; Müller and Connor 1991). These studies have led to the hypothesis that the mechanisms underlying reduction of the I_{sAHP} may occur downstream from the depolarization-induced Ca^{2+} entry. In hippocampal CA1 neurons, the I_{sAHP} seems to be activated directly by intracellular Ca^{2+} rather than by coupling through other intermediate events. As shown in excised patch recordings, single channel activities presumably underlying the I_{sAHP} exhibit fast kinetics in response to applied Ca^{2+} (Lancaster et al. 1991). A rapid, sAHP-like response is also induced in hippocampal CA1 neurons following photolytic release of caged Ca^{2+} intracellularly (Lancaster and Zucker 1994). Taken together, it is therefore reasonable to assume that the I_{sAHP} reduction might result from the a loss of sensitivity of the underlying K^+ channels to Ca^{2+}, presumably due to kinase-dependent modulation. Because the generation of the I_{sAHP} is associated with an elevation of intracellular Ca^{2+} in the nM range (Knöpfel et al. 1990; Müller and Connor 1991) and the whole-cell I_{sAHP} currents last for a few seconds, it is of interest to know whether the underlying K^+ channels can be modulated after being activated by raised intracellular Ca^{2+}.

We therefore examined whether the stimulation of cholinergic afferents is effective in reducing the I_{sAHP} when it is given immediately after the depolarizing pulse that triggers the I_{sAHP}. The afferent stimulation, when applied 1 s before the depolarizing pulse, decreased the following I_{sAHP} by 61.3±3.7% ($n=5$) in the presence of 2 μM neostigmine. In contrast, the similar stimulation, when applied 100-400 ms after the end of the pulse used to evoked the I_{sAHP}, decreased the current by only 9.8±3.8% in the same neurons (Figure 1a).

Because the I_{sAHP} reduction is mediated by receptor-coupled second messenger cascades that may require a certain time to develop, one might argue that the minor change in the I_{sAHP} seen with postdepolarization stimulation may simply be due to an insufficient latency between the afferent stimulation and the generation of the I_{sAHP}. To probe this possibility, we recorded the I_{sAHP} at room temperature (22-23°C) in an attempt to slow the energy-dependent Ca^{2+} extrusion thereby prolonging the I_{sAHP} time course. At room temperature, the duration of the I_{sAHP} was more than 10 s, which is about threefold longer than that recorded at 33°C. As with changes observed at 33°C, these I_{sAHP}s were reduced by 47±10% ($n=5$) following the afferent stimulation applied 200-300 ms before the depolarizing pulse, but showed no decrease (9.4±5%) in response to the similar stimulation applied 100-400 ms after the end of depolarizing pulses in the same neurons.

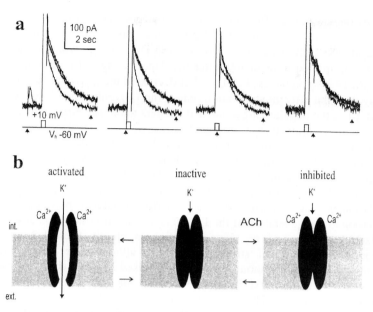

Fig. 1a,b. Proposed model for the muscarinic modulation of the Ca^{2+}-dependetn K^+ current (I_{sAHP}) that underlies the slow afterhyperpolarization (sAHP). **a** All records were collected from a hippocampal CA1 pyramidal neuron in a rat brain slice at 33°C. The neuron was whole-cell voltage-clamped at -60 mV, and constant voltage pulses to +10 mV were used to evoke the long-lasting I_{sAHP} (upward tail current following the positive pulse). Superimposed I_{sAHP} in each set was evoked before, immediately following, or 30 s after the afferent stimulation. The afferent stimulation and the subsequently evoked I_{sAHP} are indicated by *arrows*. Note that the same afferent stimulation caused a profound decrease in the following I_{sAHP} when it was given before (two panels at left), but not after (far right panel), the positive voltage pulse. **b** A hypothetical model is presented to explain the muscarinic modulation of the K^+ channels underlying the Ca^{2+}-dependent I_{sAHP}. The K^+ channels have two activity states in the absence of excessive neurotransmitter stimulation, i.e., the inactive or close state in response to basal level of intracellular Ca^{2+} (middle) and the activated or open state when stimulated by elevated intracellular Ca^{2+} (left). The acetylcholine (ACh) muscarinic inhibition takes place only when the channels are in the inactive state, leading to the inhibition state with a low open probability of the channel even though nearby Ca^{2+} signals are elevated above the basal level (right)

Prolonged I_{sAHP}s were also induced at 33°C by dialyzing hippocampal CA1 neurones with a patch pipette solution containing 1 mM BAPTA and 0.1 mM Ca^{2+}. The prolongation of I_{sAHP} by BAPTA is likely caused by redistribution of intracellular Ca^{2+} as the result of intracellular application of excessive exogenous calcium buffer with high affinity and fast kinetics (Jahromi et al 1998; Schwindt et al. 1992; Zhang et al. 1995). When stabilized in the presence of intracellular

BAPTA, the I_{sAHP} lasted for about 6 s, twice as long as control responses. When applied 200-500 ms before the depolarizing pulse, the afferent stimulation decreased the following I_{sAHP} by 66.5±5.1% ($n=5$); whereas the similar stimulation, applied immediately after the depolarizing pulse, had no significant effect (8.2±4.3%) on the evoked I_{sAHP}, but decreased the I_{sAHP} evoked 30 s later.

Why does the afferent stimulation reduce the I_{sAHP} only when it is given before, but not after, the depolarizing pulse that triggers the I_{sAHP}? First, voltage-activated Ca^{2+} currents or fluorescent Ca^{2+} signals associated with the I_{sAHP} are not affected by muscarinic agonists or the afferent stimulation (Knöpfel et al. 1990; Müller and Connor 1991; Zhang et al. 1996). Therefore, the inability to reduce I_{sAHP} by postdepolarization stimulation cannot be explained by changes in its actions on the Ca^{2+} entry. Second, the failure to decrease the I_{sAHP} by postdepolarization stimulation cannot be attributed to an insufficient latency between the afferent stimulation to the generation of the I_{sAHP}. The latency necessary to cause reduction is relatively short, and the I_{sAHP} reduction is clearly seen 1-2 s following predepolarization afferent stimulation. Third, although the postdepolarization afferent stimulation is without effect on the evoked I_{sAHP}, its ability to reduce the I_{sAHP} evoked over the next 30 s is well maintained in the presence of AChE inhibitor, neostigmine. This suggests that the afferent stimulation is capable of initiating a second-messenger cascade that is effective in reducing the I_{sAHP} once the intracellular Ca^{2+} has returned to the basal level.

From these results we suggest that the I_{sAHP} is inhibited by neurotransmitter receptor-mediated events only before its activation by the raised intracellular Ca^{2+} (Figure 1b). The activity of K^+ channel underlying the Ca^{2+}-dependent I_{sAHP} may have two main functional states in the absence of excessive neurotransmitter stimulation, i.e., inactive or activated (Valiante et al. 1997). Upon the stimulation of neurotransmitter receptors and the downstream kinase-dependent processes (Abdul-Ghani et al. 1996; Baskys et al. 1990; Müller et al. 1992; Pedarzani and Storm 1993, 1996), the inactive K^+ channels may somehow decrease their sensitivity to Ca^{2+}, therefore leading to a low channel open probability (Sah and Isaacson 1995) even though nearby Ca^{2+} is elevated above the basal level (Levitan 1994). The kinase-dependent modulation may take place on the Ca^{2+}-binding sites and/or Ca^{2+}-activating steps of the inactive K^+ channels. Occupation of these sites by Ca^{2+} may prevent any further modulation, thereby making activated channels unresponsive to the consequences of synaptic stimulation.

References

Abdul-Ghani MA, Valiante TA, Carlen PL, et al (1996) Tyrosine kinase inhibitors enhance a Ca^{2+}-activated K^+ current (I_{AHP}) and reduce I_{AHP} suppression by a metabotropic glutamate receptor agonist in rat dentate granule neurones. J Physiol 496:139-144

Baskys A, Bernstein NK, Barolet AW, et al (1990) NMDA and quisqualate reduce a Ca^{2+}-dependent K^+ current by a protein kinase-mediated mechanism. Neurosci Lett 112:76-81

Bean BP (1992) Whole-cell recording of calcium channel currents. Methods Enzymol 207:181-193

Belousov AB, Krnjevic K (1995) Internal Ca^{2+} stores involved in anoxic responses of rat hippocampal neurones. J Physiol 486:547-556

Chung I, Zhang Y, Eubanks JH, et al (1998) Attenuation of the hypoxia-induced K^+ current by intracellularly applied ATP-regenerating agents. Neurosciences 86:1101-1107

Cole AE and Nicoll RA (1983) Acetylcholine mediates a slow synaptic potential in hippocampal pyramidal cells. Science 221:1299-1301

Cole AE, Nicoll RA (1984) Characterization of a slow cholinergic post-synaptic potential recorded in vitro from rat hippocampal pyramidal cells. J Physiol 352:173-188

Dutar P, Bassant MH, Senut MC, et al (1995) The septohippocampal pathway: structure and function of central cholinergic system. Physiol Rev 75:393-427

Eagger SA, Harvey RJ (1995) Clinical heterogeneity: response to cholinergic therapy. Alzheimer Dis Assoc Disord 9 (suppl) 2:37-42

Everitt BJ, Robbins TW (1997) Central cholinergic systems and cognition. Annu Rev Psychol 48:649-684

Flentge F, Venema K, Koch T, et al (1992) An enzyme-reactor for electrochemical monitoring of choline and acetylcholine: applications in high-performance liquid chromatography, brain tissue microdialysis, and cerebrospinal fluid. Anal Biochem 204:305-310

Hsu JC, Zhang L, Wallace MC, et al (1997) Cerebral ischemia alters the regional hippocampal expression of the rat m1 muscarinic acetylcholine receptor gene. Neurosci Lett 219: 87-90.

Hulme EC, Birdsall NJM, Buckley NJ (1990) Muscarinic receptor subtypes. Annu Rev Pharmacol. Toxicol. 30:633-673.

Jahromi BS, Zhang L, Carlen PL, et al (1997) Differential time-course of slow afterhyperpolarization and associated Ca^{2+} transients in rat CA1 pyramidal neurons: further dissociation by Ca^{2+} buffer. Neurosciences 88:719-726

Kay AR (1992) An intracellular medium formulary. J Neurosci Methods 44:91-100

Knöpfel T, Vranesic I, Gähwiler BH, et al (1990) Muscarinic and β-adrenergic depression of the slow Ca^{2+} activated potassium conductance in hippocampal CA3 pyramidal cells is not mediated by a reduction of depolarization-induced cytosolic Ca^{2+} transients. Proc Natl Acad Sci USA 87:4083-4087

Krnjevic K (1993) Central cholinergic mechanisms and function. Prog. Brain Res. 98:285-92

Lancaster B, Nicoll RA, Perkel DJ (1991) Calcium activates two types of potassium channels in rat hippocampal neurons in culture. J Neurosci 11:23-30

Lancaster B, Zucker RS (1994) Photolytic manipulation of Ca^{2+} and time course of slow, Ca^{2+}-activated potassium current in rat hippocampal neurones. J Physiol 389:187-204

Levey AI, Edmunds SM, Koliatsos V, et al (1995) Expression of m1-m4 muscarinic acetylcholine receptor proteins in rat hippocampus and regulation by cholinergic innervation. J Neurosci 15:4077-4092

Levitan IB (1994) Modulation of ion channels by protein phosphorylation and dephosphorylation. Annu Rev Physiol 56:193-212

Lewis PR, Shute CCD (1967) The cholinergic limbic system: Projections to hippocampal formation, medial cortex, nuclei of the ascending cholinergic reticular system and the subformical organ and supraoptic crest. Brain 90:521-540

Madison DV, Lancaster B, Nicoll RA (1987) Voltage clamp analysis of cholinergic action in the hippocampus. J Neurosci 7:733-741

Massoulié J, Pezzementi L, Bon S, et al (1993) Molecular and cellular biology of cholinesterase. Prog Neurobiol 41:39-91

Müller W, Connor JA (1991) Cholinergic input uncouples Ca^{2+} changes from K^+ conductance activation and amplifies intradendritic Ca^{2+} changes in hippocampal neurons. Neuron 6:901-905

Müller W, Petrozzino JJ, Griffith LC, et al (1992) Specific involvement of Ca^{2+}-calmodulin kinase II in cholinergic modulation of neuronal responsiveness. J Neurophysiol 68:2264-2269

Parnetti L (1995) Clinical pharmacokinetics of drugs for Alzheimer's disease. Clin Pharmacokinet 29:110-129

Pedarzani P, Storm JF (1993) PKA mediates the effects of monoamine transmitters on the K^+ current underlying the slow spike frequency adaptation in hippocampal neurons. Neuron 13:1023-1035

Pedarzani P, Storm JF (1996) Evidence that Ca/calmodulin-dependent protein kinase mediates the modulation of the Ca^{2+}-dependent K^+ current, I_{AHP}, by acetylcholine, but not by glutamate, in hippocampal neurons. Pflügers Arch 431:7223-728

Poirier J, Delisle MC, Quirion R, et al (1995) Apolipoprotein E4 allele as a predictor of cholinergic deficits and treatment outcome in Alzheimer disease. Proc Natl Acad Sci USA 92: 12260-12264

Sah P, Isaacson JS (1995) Channels underlying the slow afterhyperpolarization in hippocampal neurons: neurotransmitters modulate the open probability. Neuron 15:435-441

Schwindt PC, Spain WJ, Crill WE (1992) Effects of intracellular calcium chelation on voltage-dependent and calcium dependent currents in cat neocortical neurons. Neurosciences 47:571-578.

Storm JF (1990) Potassium currents in hippocampal pyramidal cells. Prog Brain Res 83:161-187

Taylor L, Griffith WH (1993) Age-related decline in cholinergic synaptic transmission in hippocampus. Neurobiol. Aging 14: 509-515

Taylor P, Radic Z (1994) The cholinesterase: from genes to proteins. Annu Rev Pharmacol. Toxicol 34:281-320

Tsubokawa H, Ross WN (1996) Muscarinic modulation of spike backpropagation in the apical dendrites of hippocampal CA1 pyramidal neurons. J Neurosci 17:5782-5791

Valiante TA, Abdul-Ghani MA, Carlen PL, et al (1997) Analysis of current fluctuation during afterhyperpolarization current in dentate granule neurones of the rat hippocampus. J Physiol 499:121-134

Velumian AA, Zhang L, Carlen PL (1993) A simple method for internal perfusion of mammalian central nervous system neurons in brain slices with multiple solution changes. J Neurosci Methods 48:131-139

Velumian AA, Zhang L, Pennefather P, et al (1997) Reversible inhibition of I_K, I_{AHP}, I_h and I_{Ca} currents by internally applied gluconate in rat hippocampal neurones. Pflügers Arch 433:343-350

Wagstaff AJ, McTavish D (1994) Tacrine: a review of its pharmacodynamic and pharmacokinetic properties, and therapeutic efficacy in Alzheimer's disease. Drugs Aging 4: 510-540.

Wess J (1993) Molecular basis of muscarinic acetylcholine receptor function. Trends Pharmacol Sci 14:308-313

Whitehouse PJ, Price DL, Clark AW, et al (1981) Alzheimer disease: evidence for selective loss of cholinergic neurons in the nucleus basalis. Ann. Neurol. 10:122-126

Winkler J, Suhr ST, Grace FH, et al (1995) Essential role of acetylcholine in spatial memory. Nature 375:484-487

Wurtman RJ (1992) Choline metabolism as a basis for the selective vulnerability of cholinergic neurons. Trends Neurosci 15:117-122

Zhang L, Krnjevic K (1993) Whole-cell recording of anoxic effects on hippocampal neurons in slices. J Neurophysiol 69:118-127

Zhang L, Han D, Carlen PL (1996) Temporal specificity of muscarinic synaptic modulation of the Ca^{2+}-dependent K^+ current (I_{sAHP}) in rat hippocampal neurones. J Physiol 496:395-405

Zhang L, Weiner JL, Valiante TA, et al (1994) Whole-cell recording of the Ca^{2+}-dependent slow afterhyperpolarization in hippocampal neurons: effects of internally applied anions. Pflug. Arch. 426:247-253

Zhang L, Pennefather P, Velumian AA, et al (1995) Potentiation of a slow Ca^{2+}-dependent K^+ current (I_{sAHP}) by intracellular Ca^{2+} chelators in hippocampal CA1 neurons of rat brain slices. J Neurophysiol 74:2225-2241

Zhang L, Weiner JL, Carlen PL (1992) Muscarinic potentiation of I_K in hippocampal neurons: Electrophysiological characterization of signal transduction pathway. J Neurosci 12:4510-4520

Zhang Y, Carlen PL, Zhang L (1997) Kinetics of muscarinic reduction of I_{sAHP} in hippocampal neurons; effects of acetylcholinestrase inhibitors. J Neurophysiol 78:2999-3007

Zola-Morgan S, Squire LR (1993) Neuroanatomy and memory. Annu. Rev. Neurosciences 16:547-563

Modulation of K+ Channels in Hippocampal Neurons: Transmitters Acting via Cyclic AMP Enhance the Excitability of Hippocampal Neurons Through Kinase-Dependent and -Independent Modulation of AHP- and h-Channels

J. F. STORM[1], P. PEDARZANI[2], T. HAUG[1] and T. WINTHER[1]

[1]Department of Physiology, University of Oslo, PB 1103 Blindern, N-0317 Oslo, Norway.
[2]Max-Planck-Institute for Experimental Medicine, Department of Molecular Biology of Neuronal Signals, Hermann-Rein-Str. 3 - 37075 Göttingen, Germany

Key words: Cyclic AMP, Protein kinase A, SK Channels, h-Channels, Afterhyperpolarization, AHP current, Hippocampus, CA1, Pyramidal cells, Noradrenaline, Serotonin, Histamine, Dopamine, CRF, Corticotropin releasing factor, CGRP, Calcitonin gene related peptide, PACAP, Pituitary adenylate Cyclase-activating peptide, VIP, Vasoactive intestinal peptide.

1 Introduction

Neuronal excitability is regulated by neurotransmitters through modulation of ion channels. The channels can be modulated via two basic mechanisms: by ligand binding or by covalent modification by kinases, phosphatases and other enzymes (Hille 1992). The archetypal second messenger, cyclic AMP (cAMP) is thought primarily to act indirectly, by activating protein kinase A (PKA) which in turn can phosphorylate ion channels or their regulatory proteins. In the brain, in particular, cAMP effects have been considered to be almost exclusively PKA-mediated (Walaas and Greengard 1991). Here we review our results concerning the modulation of two types of ion channels in hippocampal pyramidal cells, indicating that monoamine transmitters acting via cAMP can exert both kinase-dependent and kinase-independent regulation of neuronal excitability in the brain.

2 Modulation of I_{sAHP} (SK channels) and Spike Frequency Adaptation

One of the most striking examples of neuromodulation in the vertebrate brain is the suppression of the slow Ca^{2+}-activated K+ current, I_{sAHP}, by noradrenaline and other monoamine transmitters (Nicoll 1988). This current underlies the slow afterhyperpolarization (sAHP) and late spike frequency adaptation in hippocampal and other cortical pyramidal cells (Sah 1996). It is most likely mediated by small-conductance (SK) Ca^{2+}-activated K^+ channels. Recently, the first SK channels were cloned, and hippocampal pyramidal cells were found to express a gene (SK1) encoding apamin-insensitive SK-channels (Köhler et al. 1996) that probably underlie the apamin-insensitive hippocampal sAHP (Lancaster and Adams 1996; Storm 1990; Sah 1996).

The suppression of I_{sAHP} in cortical pyramidal cells is a major effector mechanism of the ascending monoaminergic fiber systems: noradrenergic projections from locus coeruleus, serotonergic projections from the raphe nuclei, dopaminergic projections from the ventral tegmental area, and histaminergic fibers from the hypothalamic mamillary region (Verney et al 1985; Moore and Bloom 1979; Pollard and Schwartz 1987; Azmitia 1987; Nicoll 1988; Nicoll et al. 1990). These wide projections control the functional state of the entire forebrain, leading to shifts from sleep to wakefulness, attention, arousal, as well as modulation of sensory perception, emotions and cognitive functions (Chu and Bloom 1973; Foote et al 1980; Aston-Jones and Bloom 1981; Livingstone and Hubel 1981; McCormick 1989; Steriade and McCarley 1990).

The effect of noradrenaline on I_{sAHP} in hippocampal pyramidal neurons is mediated by β_1 receptors via adenylyl cyclase and cyclic AMP (cAMP) (Madison and Nicoll 1982, 1986a,b; Haas and Konnerth 1983). cAMP also mediates the effects of serotonin, via 5-HT$_4$-like receptors (Dumuis et al 1988; Chaput et al 1990; Andrade and Chaput 1991; Zifa and Fillion 1992), and of histamine, via H$_2$ receptors (Haas 1985; Haas and Greene 1986).

Protein kinase A is regarded as the main intracellular receptor protein for cAMP, and it was proposed to mediate the cAMP-induced modulation of I_{sAHP} (Nicoll 1988; Nicoll et al 1990). However, in a variety of cell types, including heart muscle, photoreceptors, olfactory receptors, and *Drosophila* muscle, cAMP and other cyclic nucleotides can modulate ion channels directly, in a kinase-independent manner (Fesenko et al. 1985; Nakamura and Gold 1987; Delgado et al. 1991; DiFrancesco and Tortora 1991), and it was proposed that a kinase-independent mechanism mediates suppression of the sAHP by noradrenaline and cAMP in locus coeruleus neurons (Aston-Jones and Shiekhattar 1992). On this background, we reexamined the hypothesis that the suppression of I_{sAHP} by monoaminergic transmitters and cAMP in pyramidal cells is mediated by PKA (Pedarzani and Storm 1993, 1995a). Our results provide three lines of evidence that protein phosphorylation by PKA mediates the modulation of I_{sAHP} by noradrenaline, serotonin, dopamine and histamine in hippocampal pyramidal cells.

2.1 Whole-Cell Recording of I_{sAHP} and Its Modulation

The sAHP-currents (I_{sAHP}) was measured by whole-cell recording from CA1 pyramidal cells at room temperature in hippocampal slices from young rats, using a K gluconate-based intracellular medium. Most cells showed a typical I_{sAHP}, i.e., a slow tail current following a train of action potentials or a Ca^{2+} spike. The I_{sAHP} remained sensitive to transmitter substances and cAMP analogues for the duration of the recordings (Pedarzani and Storm 1993).

2.2 Protein Kinase A Mediates the Effect of Noradrenaline on I_{sAHP} and Spike Frequency Adaptation

2.2.1 The Effects of β–Receptor Agonists and cAMP Analogues Were Prevented by PKA Inhibitors and Mimicked by PKA Injection

To study the mechanism of suppression of I_{sAHP} via β1 adrenegic receptors, we used the β–receptor agonist isoproterenol (isoprenaline). The involvement of PKA was tested with two PKA inhibitors, which block the kinase in different ways: Rp-cAMPS, which blocks the cAMP-binding sites on the regulatory subunits of PKA (Bothelo et al 1988), and Walsh peptide (PKI; Cheng et al. 1986), which blocks the catalytic site. Each of these inhibitors substantially reduced the suppression of I_{sAHP} by isoprenaline, noradrenaline and cAMP analogues, indicating that the β–receptor effect is mediated by PKA. This kinase is the only signaling molecule known to be sensitive to both Rp-cAMPS and PKI. Furthermore, intracellular application of PKA catalytic subunit (PKA-C) mimicked the effects of noradrenaline and cAMP, causing a gradual decline of I_{sAHP} (Pedarzani and Storm 1993). These results strongly suggest that PKA mediates the effects of noradrenaline, isoproterenol, and cAMP on I_{sAHP}.

2.2.2 Inhibition of PKA Prevented the Effects of Noradrenaline on Spike Frequency Adaptation and the Slow AHP

Suppression of I_{sAHP} by transmitters reduces spike-frequency adaptation and increases the number of action potentials in response to current injection (Nicoll 1988). To study the functional effects of PKA inhibitors on spike frequency modulation, and to exclude age dependence or effects of cell dialysis or temperature, we also used intracellular recording with sharp microelectrodes in hippocampal slices from adult rats at 30°-$32^{\circ}C$. We found that the characteristic effects of noradrenaline, reduction in spike frequency adaptation and inhibition of the sAHP, were largely suppressed by Rp-cAMPS in the electrode, thus confirming our conclusions from the whole-cell recordings.

2.3 Serotonin-, Histamine-, and Dopamine-Induced Modulation of I_{sAHP} and Spike Frequency Also Depend on PKA

We also tested the modulation of I_{sAHP} by five other neurotransmitters: serotonin, histamine, dopamine, acetylcholine and glutamate. The PKA inhibitor Rp-cAMPS was found to prevent the effects of serotonin, histamine, and dopamine, both in whole-cell voltage clamp recordings and in current clamp recordings of spike frequency adaptation and the sAHP. In contrast, the effects on I_{sAHP} of the cholinergic agonist carbachol or the metabotropic glutamate receptor (mGluR) agonist trans-ACPD were not inhibited by Rp-cAMPS, even in cells where this PKA inhibitor effectively suppressed the effects of monoamine transmitters (Pedarzani and Storm 1993, 1996a; Blitzer et al. 1994). These results strongly suggest that all the four monoamine transmitters act via PKA, whereas muscarinic acetylcholine and metabotropic glutamate receptors use other signal pathways (Gerber et al. 1992; Sim et al. 1992; Müller et al. 1992; Pedarzani and Storm 1996a; Krause et al. 1997).

2.4 Synergism Between α-and β-Adrenergic Receptor Agonists in PKA-Dependent Modulation of I_{sAHP}

In addition to the purely β-adrenergic effect on I_{sAHP} (Madison & Nicoll 1982), we found evidence for a strong synergistic interaction between β-adrenergic receptors and receptors for the α-adrenergic agonist 6-fluoro-norepinephrine (6Fl-NE). The latter receptors are probably α-adrenergic, but show a nonclassical pharmacological profile (Pedarzani and Storm 1996b). Thus, combined application of both isoproterenol and 6Fl-NE (Cantacuzene et al. 1979; Daly et al. 1980; Chiueh et al. 1983; Adejare et al. 1988), produced a far stronger suppression of I_{sAHP} than either agonist alone, and our data are compatible with an interaction at the G-protein level (Pedarzani and Storm 1996b). This synergistic interaction is in accordance with biochemical data showing an enhanced stimulation of cAMP production in the brain, by cross-talk between α and β adrenergic receptors (Perkins and Moore 1973; Daly et al. 1981; Leblanc and Ciaranello 1984; Stone and Herrera 1986; Duman and Enna 1987; Robinson and Kendall 1989; Atkinson and Minneman 1991). Thus, both receptor types may contribute to the increase in excitability and enhancement of the signal-to-noise ratio, which are thought to result from noradrenergic action on cortical neurons (Woodward et al 1979; Madison and Nicoll 1986a; Pedarzani and Storm 1996b). In addition to the short-term modulation of I_{sAHP} measured in our study, the α–β synergism may also contribute to long-term cAMP-dependent regulation. It is known that high levels of cAMP, such as those resulting from the synergistic stimulation of α- and β-adrenoceptors, may cause translocation of PKA catalytic subunit to the nucleus (Bacskai et al. 1993), and influence gene expression via CREB protein and CRE (Lee and Masson 1993; Stevens 1994). This may lead to long-lasting effects such as long-term synaptic plasticity (Goelet and Kandel 1986; Bacskai et al. 1993; Frey et al. 1993; Stevens 1994).

In general, similar cross-talk mechanisms, caused for example by the convergence of $G_{\alpha s}$ and $\beta\gamma$ subunits from G_i/G_o on type II adenylyl cyclase, can participate in integration between different transmitters and second messenger pathways (Tang and Gilman 1991; Hille 1992; Andrade 1993; Bourne and Nicoll 1993; Clapham and Neer 1993). For example, noradrenaline acting via α-adrenergic receptors may enhance the effect of other transmitters stimulating G_s-coupled receptors, such as serotonin acting on 5-HT4 receptors or histamine on H_2 receptors. Conversely, other transmitters stimulating G_i/G_o-coupled receptors may potentiate the effect of β-receptor stimulation by noradrenaline (Andrade 1993; Gereau and Conn 1994). Andrade (1993) proposed that this mechanism underlies the potentiation by $GABA_B$ and $5-HT_{1A}$ receptors of the β-adrenergic suppression of I_{sAHP} in the hippocampus, and α-adrenergic receptors may also act through this mechanism (Tang and Gilman 1991). Thus, cross-talk between different receptors may be used for integration between different neurotransmitters and function as a coincidence detector (Bourne and Nicoll 1993).

2.5 Peptidergic Modulation of I_{sAHP}: CGRP, CRF, PACAP and VIP Ssuppress I_{sAHP} via cAMP and PKA

In addition to the widespread actions of the ascending monoamine activation systems acting via cAMP and PKA, several neuropeptide transmitters supplied by interneurons may also modulate excitability in the forebrain in a more local manner, using similar intracellular signaling pathways (Magistretti and Morrison 1985). The neuropeptides CRF (corticotropin releasing factor), CGRP (calcitonin gene related peptide), PACAP (pituitary adenylate cyclase-activating peptide), and VIP (vasoactive intestinal peptide) are known to stimulate production of cyclic AMP (cAMP) in the brain (Etgen and Browning 1983, Magistretti and Schorderet 1984, Goltzman and Mitchel 1985, Eusebi et al. 1987), and CFR and VIP have been shown to suppress I_{sAHP} in hippocampal neurons (Aldenhoff et al. 1983; Haas and Gähwiler 1992).

Recent experiments indicate that at least some of these peptidergic actions are also mediated by PKA in CA1 pyramidal cells (Haug and Storm 1998). We found that I_{sAHP} was suppressed by CRF, CGRP, VIP, and PACAP-38 under normal conditions, but when the PKA inhibitor Rp-cAMPS was added to the intracellular medium, CRF, CGRP, and VIP had little or no effect on I_{sAHP}. These results suggest that PKA mediates the modulation of I_{sAHP} by CGRP, CRF, and VIP.

2.6 Evidence for a Basal Regulation of I_{sAHP} by a Phosphatase-Kinase Balance

If protein kinases, such as PKA, have a certain basal activity even in unstimulated cells, a continuous action of protein phosphatases may be needed to avoid a persistent suppression of I_{sAHP}. If so, inhibition of protein phosphatases should lead to a rundown in I_{sAHP} in the absence of agonists that stimulate phosphorylation. This hypothesis was tested by adding different broad spectrum-serine/threonine phosphatase inhibitors (microcystin-LR, calyculin A, or cantharidic acid) to the intracellular medium. All these phosphatase inhibitors produced a gradual decrease of I_{sAHP} amplitude (Pedarzani and Storm 1993; Pedarzani et al. 1998). In contrast, cells recorded without phosphatase inhibitors in the pipette showed no decline in I_{sAHP}. These results further support the role of protein phosphorylation in the modulation of I_{sAHP}, and suggest the presence of an ongoing phosphorylation/dephosphorylation turnover, due to a basal kinase activity that is balanced by phosphatases, even in the absence of applied transmitters.

To test which phosphatase is involved in this basal modulation of I_{sAHP}, we used selective inhibitors. FK-506 and calcineurin autoinhibitory peptide, which inhibit the Ca^{2+}/calmodulin-dependent phosphatase 2B (PP-2B, also called calcineurin; Hashimoto et al. 1990), did not affect the amplitude of IsAHP (Pedarzani et al.1998). This result leaves protein phosphatase 1 (PP-1) and 2A (PP-2A) as the phosphatases most likely responsible for the basal modulation of I_{sAHP}.

To identify the protein kinase causing the basal phosphorylation, we coapplied the PKA inhibitor Rp-cAMPS and the phosphatase inhibitor microcystin through the patch pipette. During this combined application, no decrease in the I_{sAHP} amplitude was observed, in sharp contrast to the fast decline caused by microcystin alone (Pedarzani et al. 1998). This result suggests that PKA is mainly responsible for the steady-state modulation of I_{sAHP} under basal conditions. This conclusion was further supported by the following results: (1) PKA inhibition by Rp-cAMPS enhanced I_{sAHP} under basal conditions (Gerber and Gähwiler 1994; Pedarzani et al. 1998), (2) an adenylyl cyclase inhibitor enhanced I_{sAHP}, and (3) phosphodiesterase inhibitors reduced I_{sAHP} (Madison and Nicoll 1986b; Pedarzani et al. 1998). Thus, a tonic activity of adenylyl cyclase seems to maintain a cAMP level sufficient to tonically inhibit I_{sAHP} under basal conditions. The balance between phosphorylation (by PKA) and dephosphorylation (by PP-1 or PP-2A) constitutes a "cyclic cascade" that may exert a dynamic and fine-tuned regulation of I_{sAHP} and repetitive firing properties (Shacter et al. 1988). Little is known about the localization of AHP channels relative to the receptors, G-proteins, adenylyl cyclase, kinases, and phosphatases. If these molecules are colocalized in functional domains (Reinhart and Levitan 1995; Faux and Scott 1996), this may substantially speed up the responses to stimuli (Shacter et al. 1988).

3. Modulation of H-current via a Kinase-Independent Effect of Cyclic AMP

Transmitters acting via cAMP usually modulate target proteins by PKA activation and phosphorylation, which is considered to be the main mechanism of cAMP action in the brain (Walaas and Greengard 1991). Nevertheless, it was recently shown that cAMP also act in a kinase-independent manner to modulate neuronal excitibility in the brain, by modulation of h-channels (Pedarzani and Storm 1995b). Here we will review the evidence for this second branch of the cAMP pathway in hippocampal pyramidal cells.

3.1 A PKA-Independent Depolarization of Hippocampal Neurons by cAMP

When applying cAMP analogues or β-adrenergic receptor agonists to study I_{sAHP} modulation, we noticed a concomitant inward resting current, associated with an increase in membrane conductance (Pedarzani and Storm 1995b). Similarly, in current clamp, cAMP- or β-agonists caused a moderate membrane depolarization (~4-10 mV), which could elicit action potentials. The depolarization or inward current was blocked by the β-receptor antagonists and occluded by cAMP in the patch pipette, indicating that it was due to a β-receptor-induced rise in $[cAMP]_i$, just like the suppression of I_{sAHP} (Pedarzani and Storm 1993). Nevertheless, when PKA was blocked by PKI (Walsh peptide), there was a striking difference in the effects on the two actions of isoprenaline. The inward shift in the holding current was not diminished in cells loaded with PKI, whereas the suppression of I_{sAHP} by isoprenaline was largely prevented in these cells (Pedarzani and Storm 1995b). Since the β agonist-evoked inward current was fully occluded by cAMP but not prevented by blockade of PKA, we conclude that it must be caused by a PKA-independent modulation by cAMP.

3.2 PKA-Independent cAMP-Mediated Enhancement of the h-Current Underlies the β-adrenergic Depolarization

To identify the channel type underlying the cAMP-induced inward current, we applied a variety of Na^+, K^+, and Ca^{2+} channel blockers. TTX, TEA, 4-AP, Ba^{2+}, and Mn^{2+} all failed to reduce the cAMP-induced inward current, whereas it was fully blocked by 1 mM Cs^+. This pharmacological profile points unequivocally to the h-current (I_h, also called Q-current, I_Q), a mixed Na^+/K^+ inward current that is slowly activated by hyperpolarization (Halliwell and Adams 1982; Pape 1996). This conclusion was further supported by the observations that the isoprenaline-induced inward current was seen only at potentials negative to –60 mV, corresponding to the activation range of I_h, and that I_h evoked by hyperpolarizing voltage steps was clearly enhanced by isoprenaline or the cAMP analogue 8CPT-cAMP. Isoprenaline appeared to induce a rightward shift in the activation curve of I_h, rendering it more activated at the resting potential. The enhanced I_h also showed a faster activation at all voltages. A similar voltage shift and change in kinetics by cAMP has previously been shown for h currents in other cell types, including heart muscle and thalamic relay cells (DiFrancesco and Tortora 1991; Pape and McCormick 1989).

We conclude that β-adrenergic receptor agonists depolarize CA1 hippocampal pyramidal cells by enhancing the h-channel activity at the resting potential. Our data indicate that this action is mediated by cAMP, but not via PKA, indicating for the first time a kinase-independent modulatory effect of cAMP in the brain. A similar mechanism was previously shown in the heart SA-node where sympathetic nerves and adrenaline accelerate the heartbeat via direct cAMP enhancement of h-type channel activity (DiFrancesco and Tortora 1991). Thus, there seems to be a striking parallel between the molecular mechanisms operating in the brain and in the heart during arousal.

3.3 Kinase-Independent cAMP-Mediated Modulation of the h-Current by Other Monoamine Transmitters: Serotonin, Histamine and Dopamine

We next tested the effects of other monoamine transmitters on the h-current. Serotonin, histamine, and dopamine enhanced I_h in a manner similar to isoprenaline or 8CPT-cAMP and to monoamine effects on I_h in other CNS neurons (Pape and McCormick 1989; Ingram and Williams 1996). Thus, it appears that in CA1 pyramidal cells all the monoamine transmitters that activate cAMP production modulate both I_{sAHP} and I_h in parallel. These results suggest a convergent modulation of I_h by several transmitters involved in brain state control and arousal, most likely through a direct cyclic nucleotide effect on the h-channels. Because the effect of phosphodiesterase inhibitors on I_{sAHP} indicate a basal cAMP production even in unstimulated cells (Madison and Nicoll 1986b; Pedarzani et al. 1998), it seems likely that I_h is also subject to basal modulation.

3.4 The PKA-Independent cAMP-Mediated Modulation of the h-Current is Occluded by the PKA-Blocking cAMP Analogue Rp-cAMPS

The cAMP analogue Rp-cAMPS is widely used as a PKA inhibitor (Bothelo et al. 1988). In our studies of the sAHP current, we found that the effects of Rp-cAMPS were very similar to those of the pseudosubstrate peptide inhibitor PKI, since both PKA inhibitors prevented the modulation of sAHP by monoamines and cAMP (Pedarzani and Storm 1993). In contrast, when tested on the h-current in the same CA1 cells, we found that the two inhibitors had very different effects (Storm et al.

1996). PKI had no measurable effect on I_h or its modulation, whereas Rp-cAMPS apparently prevented the enhancement of I_h by isoprenaline, serotonin, histamine, or dopamine. It seems unlikely that this effect was due to blockade of PKA, since PKI had no effect on I_h modulation despite efficiently blocking the PKA-dependent modulation of I_{sAHP} in the same cells (Pedarzani and Storm 1995b). Instead, Rp-cAMPS may bind to the cAMP-binding sites not only on PKA, but also on other proteins including the h-channels themselves, thus blocking or mimicking the effects of cAMP. To test this hypothesis, we compared the amplitudes of I_h in cells recorded with and without Rp-cAMPS in the pipette, and found that I_h was significantly larger in cells loaded with Rp-cAMPS (Storm et al. 1996). This suggests that Rp-cAMPS activates the same binding site as cAMP, probably at the h-channels themselves, thus enhancing I_h and occluding the effects of cAMP and monoamines. Similar effects of Rp-cAMPS have been described for primary afferent ganglion neurons (Ingram and Williams 1996) and for the cyclic nucleotide-gated channels in retinal photoreceptors (Kramer and Tibbs 1996).

3.5 Comparison of Hippocampal h-Channel Modulation with Results from Other Cell Types and Cloned h-Channels

We concluded that norepinephrine and other monoamine transmitters induce a cAMP-mediated but protein kinase A-independent shift in the voltage dependence of I_h in CA1 pyramidal cells. A similar kinase-independent modulation of I_h is likely to exist in the thalamus (Pape and McCormick 1989) and other brain regions where modulation of I_h is thought to be important for state control and to participate in the activation of different parts of the forebrain during arousal and attention (McCormick 1989; Steriade and McCarley 1990; Pape 1996; Lüthi and McCormick 1998).

Recently, the first h-channel genes have been cloned: mBCNG-1 (HAC-2) (Santoro et al. 1997, 1998), HAC1 (BCNG-2) (Gauss et al. 1998), and SPIH (Ludwig et al. 1998). These channel genes belong to a new gene family related to the voltage-gated K^+ channels, the cyclic nucleotide-gated nonselective cation channels and the plant inwardly rectifying K^+ channel KAT1 (for a review, see Clapham 1998). The deduced structures of the cloned h-channels all show a cyclic nucleotide-binding region, and cAMP induces a 2-12 mV shift in the activation curve of the expressed channels. Thus, the kinase-independent cAMP-sensitivity of the h-current can now be explained and further investigated at the molecular level.

4. Summary and Conclusions

The data reviewed here indicate that a variety of monoamine and peptide transmitters converge on the cAMP pathway, which divides into two branches (Fig. 1), both enhancing the excitability of CA1 hippocampal pyramidal cells: (1) cAMP activates A-kinase which, through phosphorylation, suppresses the SK channels underlying the slow afterhyperpolarization and spike frequency adaptation. (2) In parallel, cAMP directly, in a kinase-independent manner, stimulates the h channels, thus causing them to open more and depolarize the cell. There is little reason to doubt that the rise in cAMP caused by the peptide transmitters can also modulate I_h.

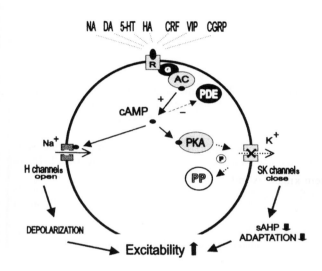

Fig. 1. Schematic representation of our working hypothesis for cyclic AMP (cAMP) modulation of AHP channels (SK) and h channels (H) in CA1 hippocampal pyramidal cells. Several different monoamine and peptide transmitter substances (NA, DA, 5-HT, HA, CRF, VIP, CGRP) activate different receptors (R), which are here represented by a single receptor molecule (R) for simplicity. The receptors activate Gs protein (G) and adenylyl cyclase (AC) thus enhancing cAMP production. cAMP modulates two types of ion channels: (1) cAMP binds to the h channels underlying I_h, causing them to open more, thus depolarizing the cell. (2) cAMP also activates protein kinase A (PKA) which, via phosphorylation (P) of an unknown target protein, closes the SK-type K^+ channels underlying I_{sAHP}, thus suppressing the slow afterhyperpolarization (sAHP) and spike frequency adaptation. These two branches of the cAMP-dependent signal pathway thus cooperate to increase the excitability of the cell. Phosphatases (PP) reverse the phosphorylation (P) caused by PKA. At rest in unstimulated cells there is an ongoing basal phosphorylation/dephosphorylation turnover by PKA and PP, keeping the SK channels at a steady submaximal level of activity. Not included in this scheme is the cross-talk between G_s-coupled receptors and other transmitter receptors (see the text).

The transmitter pathways utilizing these two excitatory cAMP-dependent mechanisms belong to two main groups: (1) the long and widely branched monoaminergic axon projections from the brainstem and basal forebrain (with noradrenaline, serotonin, histamine, dopamine) mediating coordinated state shifts in large parts of the forebrain, and (2) the peptidergic input from local interneurons (with CRF, VIP, CGRP), which is better suited for local and detailed fine control of single cells and small circuits. Thus, the two forms of cAMP-mediated neuromodulation discussed here are likely to be involved in both local and global regulation of brain activity.

ACKNOWLEDGMENTS. This work was supported by the Norwegian Research Council (NFR), the Nansen, Langfeldt and Jahre Foundations, the Norwegain Odd Fellow Foundation and by a HFSP Long-term Fellowship to P.P. We thank Dr. Martin Kelly for comments on the manuscript.

References

Adejare A, Gusovsky F, Padgett W, et al (1988) Syntheses and adrenergic activities of ring-fluorinated epinephrines. J Med Chem 31:1972-1977

Aldenhoff J B, Gruol D L, Rivier J, et al (1983) Corticotropin releasing factor decreases postburst hyperpolarizations and excites hippocampal neurons. Science 221:875-877

Andrade R (1993) Enhancement of beta-adrenergic responses by G_i-linked receptors in rat hippocampus. Neuron 10:83-88

Andrade R, Chaput Y (1991) 5-Hydroxytryptamine-like receptors mediate the slow excitatory response to serotonin in the rat hippocampus. J Pharmacol Exp Ther 257:930-937

Aston-Jones G, Bloom F E (1981) Activity of norepinephrine-containing locus coeruleus neurons in behaving rats anticipates fluctuations in the sleep-waking cycle. J Neurosci 1:876-886

Aston-Jones G, Schiekhattar R (1992) Attenuation of after hyperpolarization in locus coeruleus neurons by cAMP independent of protein kinase activation. Soc Neurosci Abstr 18:103

Atkinson B N, Minneman K P (1991) Multiple adrenergic receptor subtypes controlling cyclic AMP formation: comparison of brain slices and primary neuronal and glial cultures. J Neurochem 56:587-595

Azmitia E C (1987) The CNS serotoninergic system: progression toward a collaborative organization. In Psychopharmacology: The Third Generation of Progress, H Y Meltzer ed, Raven Press, New York, 61-74

Bacskai B J, Hochner B, Mahaut Smith M, et al (1993) Spatially resolved dynamics of cAMP and protein kinase A subunits in Aplysia sensory neurons. Science 260:222-226

Blitzer R D, Wong T, Nouranifar R, et al (1994) The cholinergic inhibition of afterhyperpolarization in rat hippocampus is independent of cAMP-dependent protein kinase. Brain Res 646:312-314

Botelho L H, Rothermel J D, Coombs R V, et al (1988) cAMP analog antagonists of cAMP action. Meth Enzymol 159:159-172

Bourne H R, Nicoll R (1993) Molecular machines integrate coincident synaptic signals. Cell 72:65-75

Cantacuzene D, Kirk K L, McCulloh D H, et al (1979) Effect of fluorine substitution on the agonist specificity of norepinephrine. Science 204:1217-1219

Chaput Y, Araneda R C, Andrade R (1990) Pharmacological and functional analysis of a novel serotonin receptor in the rat hippocampus. Eur J Pharmacol 182:441-456

Cheng H C, Kemp B E, Pearson R B, et al (1986) A potent synthetic peptide inhibitor of the cAMP-dependent protein kinase. J Biol Chem 261:989-992

Chiueh C C, Zukowska-Grojec Z, Kirk K L, et al (1983) 6-Fluorocatecholamines as false adrenergic neurotransmitters. J Pharmacol Exp Ther 225:529-533

Chu N, Bloom F E (1973) Norepinephrine-containing neurons: changes in spontaneous discharge patterns during sleep and waking. Science 179:908-910

Clapham D E (1998) Not so funny anymore. Pacing channels are cloned. Neuron 21:5-7

Clapham D E, Neer E J (1993) New roles for G-protein beta gamma-dimers in transmembrane signalling. Nature 365:403-406

Daly J W, Padgett W, Nimitkitpaisan Y, et al (1980) Fluoronorepinephrines: specific agonists for the activation of alpha and beta adrenergic-sensitive cyclic AMP-generating systems in brain slices. J Pharmacol Exp Ther 212:382-389

Daly J W, Padgett W, Creveling C R, et al (1981) Cyclic AMP-generating systems: regional differences in activation by adrenergic receptors in rat brain. J Neurosci 1:49-59

Delgado R, Hidalgo P, Diaz F, et al (1991) A cyclic AMP-activated K^+ channel in Drosophila larval muscle is persistently activated in dunce. Proc Natl Acad Sci U S A 88:557-560

DiFrancesco D, Tortora P (1991) Direct activation of cardiac pacemaker channels by intracellular cyclic AMP. Nature 351:145-147

Duman R S, Enna S J (1987) Modulation of receptor-mediated cyclic AMP production in brain. Neuropharmacology 26:981-986

Dumuis A, Bouhelal R, Sebben M, et al (1988). A nonclassical 5-hydroxytryptamine receptor positively coupled with adenylate cyclase in the central nervous system. Mol Pharmacol 34:880-887

Etgen A M, Browning E T (1983) Activators of cyclic adenosine 3':5'-monophosphate accumulation in rat hippocampal slices: action of vasoactive intestinal peptide (VIP). J Neurosci 3:2487-2493

Eusebi F, Farini D, Grassi F, Ruzzier F (1988) Effects of calcitonin gene-related peptide on synaptic acetylcholine receptor-channels in rat muscle fibers. Proc Royal Soc London 234:333-342

Faux M C, Scott J D (1996) More on target with protein phosphorylation: conferring specificity by location. Trends Biochem Sci 21:312-315

Fesenko E E, Kolesnikov S S, Lyubarsky A L (1985) Induction by cyclic GMP of cationic conductance in plasma membrane of retinal rod outer segment. Nature 313:310-313

Foote S L, Aston-Jones G, Bloom F E (1980) Impulse activity of locus coeruleus neurons in awake rats and monkeys is a function of sensory stimulation and arousal. Proc Natl Acad Sci U S A 77:3033-3037

Frey U, Huang Y Y, Kandel E R (1993) Effects of cAMP simulate a late stage of LTP in hippocampal CA1 neurons. Science 260:1661-1664

Gauss R, Seifert R, Kaupp U B (1998) Molecular identification of a hyperpolarization-activated channel in sea urchin sperm. Nature 393:583-587

Gerber U, Gähwiler B H (1994) $GABA_B$ and adenosine receptors mediate enhancement of the K^+ current, I_{AHP}, by reducing adenylyl cyclase activity in rat CA3 hippocampal neurons. J Neurophysiol 72:2360-2367

Gerber U, Sim J A, Gähwiler B H (1992) Reduction of potassium conductances mediated by metabotropic glutamate receptors in rat CA3 pyramidal cells does not require protein kinase C or protein kinase A. Eur J Neurosci 4:792-797

Gereau R W, Conn P J (1994) A cyclic AMP-dependent form of associative synaptic plasticity induced by coactivation of beta-adrenergic receptors and metabotropic glutamate receptors in rat hippocampus. J Neurosci 14:3310-3318

Goelet P, Kandel E R (1986) Tracking the flow of learned information from membrane receptors to genome. Trends Neurosci 10:492-499

Goltzmann D, Mitchell J (1985) Interaction of calcitonin gene-related peptide at receptor sites in target tissues. Science 227:1343-1345

Haas H L (1985) Histamine may act through cyclic AMP on hippocampal neurones. Agents Actions 16:234-235

Haas H L, Gähwiler B H (1992) Vasoactive intestinal polypeptide modulates neuronal excitability in hippocampal slices of the rat. Neuroscience 47:273-277

Haas H L, Greene R W (1986) Effects of histamine on hippocampal pyramidal cells of the rat in vitro. Exp Brain Res 62: 123-130

Haas H L, Konnerth A (1983) Histamine and noradrenaline decrease calcium-activated potassium conductance in hippocampal pyramidal cells. Nature 302:432-434

Halliwell J V, Adams P R (1982) Voltage-clamp analysis of muscarinic excitation in hippocampal neurons. Brain Res 250:71-92

Hashimoto Y, Perrino B A, Soderling T R (1990) Identification of an autoinhibitory domain in calcineurin. J Biol Chem 265:1924-1927

Haug T, Storm J F (1998) Peptidergic modulation via protein kinase A (PKA). of the slow Ca^{2+}-dependent K^+ current I_{AHP} in hippocampal pyramidal cells. Soc Neurosci Abstr 24: 1334.

Hille B (1992) G protein-coupled mechanisms and nervous signaling. Neuron 9:187-195

Ingram S L, Williams J T (1996) Modulation of the hyperpolarization-activated current (I_h) by cyclic nucleotides in guinea-pig primary afferent neurons. J Physiol Lond 492:97-106

Köhler M, Hirschberg B, Bond C T, et al (1996) Small-conductance, calcium-activated potassium channels from mammalian brain. Science 273:1709-1714

Kramer R H, Tibbs G R (1996) Antagonists of cyclic nucleotide-gated channels and molecular mapping of their site of action. J Neurosci 16:1285-1293

Krause M, Stühmer W, Pedarzani P (1997) Involvement of a protein phosphatase in the cholinergic suppression of the Ca^{2+}-activated K^+ current sI_{AHP} in CA1 pyramidal neurons. Pflügers Arch Suppl 434:R100

Lancaster B, Adams P R (1986) Calcium-dependent current generating the afterhyperpolarization of hippocampal neurons. J Neurophysiol 55:1268-1282

Leblanc G G, Ciaranello R D (1984) α-Noradrenergic potentiation of neurotransmitter-stimulated cAMP production in rat striatal slices. Brain Res 293:57-65

Lee K A, Masson N (1993) Transcriptional regulation by CREB and its relatives. Biochim Biophys Acta 1174:221-233

Livingstone M S, Hubel D H (1981) Effects of sleep and arousal on the processing of visual information in the cat. Nature 291:554-561

Ludwig A, Zong X G, Jeglitsch M, et al (1998) A family of hyperpolarization-activated mammalian cation channels. Nature 393:587-591

Lüthi A, McCormick D A (1998) H-current: properties of a neuronal and network pacemaker. Neuron 21:9-12

Madison D V, Nicoll R A (1982) Noradrenaline blocks accommodation of pyramidal cell discharge in the hippocampus. Nature 299:636-638

Madison D V, Nicoll R A (1986a) Actions of noradrenaline recorded intracellularly in rat hippocampal CA1 pyramidal neurones, in vitro. J Physiol Lond 372:221-244

Madison D V, Nicoll R A (1986b) Cyclic adenosine 3',5'-monophosphate mediates beta-receptor actions of noradrenaline in rat hippocampal pyramidal cells. J Physiol Lond 372:245-259

Magistretti P J, Schorderet M (1984) VIP and noradrenaline act synergistically to increase cyclic AMP in cerebral cortex. Nature 308:280-282

Magistretti P J, Morrison J H (1985) VIP neurons in the neocortex. Trends Neurosci ??:7-8

McCormick D A (1989) Cholinergic and noradrenergic modulation of thalamocortical processing. Trends Neurosci 12:215-221

Moore R Y, Bloom F E (1979) Central catecholamine neuron systems: anatomy and physiology of the norepinephrine and epinephrine systems. Ann Rev Neurosci 2:113-168

Müller W, Petrozzino J J, Griffith L C, et al (1992) Specific involvement of Ca^{2+}-calmodulin kinase II in cholinergic modulation of neuronal responsiveness. J Neurophysiol 68:2264-2269

Nakamura T, Gold G H (1987) A cyclic nucleotide-gated conductance in olfactory receptor cilia. Nature 325:442-444

Nicoll R A (1988) The coupling of neurotransmitter receptors to ion channels in the brain. Science 241:545-551

Nicoll R A, Malenka R C, Kauer J A (1990) Functional comparison of neurotransmitter receptor subtypes in mammalian central nervous system. Physiol Rev 70:513-565

Pape H C (1996) Queer current and pacemaker - The hyperpolarization-activated cation current in neurons. Ann Rev Physiol 58:299-327

Pape H C, McCormick D A (1989) Noradrenaline and serotonin selectively modulate thalamic burst firing by enhancing a hyperpolarization-activated cation current. Nature 340:715-718

Pedarzani P, Storm J F (1993) PKA mediates the effects of monoamine transmitters on the K^+ current underlying the slow spike frequency adaptation in hippocampal neurons. Neuron 11:1023-1035

Pedarzani P, Storm J F (1995a) Dopamine modulates the slow Ca^{2+}-activated K^+ current I_{AHP} via cyclic AMP-dependent protein kinase in hippocampal neurons. J Neurophysiol 74:2749-2753

Pedarzani P, Storm J F (1995b) Protein kinase A-independent modulation of ion channels in the brain by cyclic AMP. Proc Natl Acad Sci U S A 92:11716-11720

Pedarzani P, Storm J F (1996a) Evidence that Ca/calmodulin-dependent protein kinase mediates the modulation of the Ca^{2+}-dependent K^+ current, I_{AHP}, by acetylcholine, but not by glutamate, in hippocampal neurons. Pflügers Arch 431:723-728

Pedarzani P, Storm J F (1996b) Interaction between alpha- and beta-adrenergic receptor agonists modulating the slow Ca^{2+}-activated K^+ current I_{AHP} in hippocampal neurons. Eur J Neurosci 8:2098-2110

Pedarzani P, Krause M, Haug T, et al (1998) Modulation of the Ca^{2+}-activated K^+ current sI_{AHP} by a phosphatase-kinase balance under basal conditions in rat CA1 pyramidal neurons. J Neurophysiol 79:3252-3256

Perkins J P, Moore M M (1973) Characterization of the adrenergic receptors mediating a rise in cyclic 3'-5'-adenosine monophosphate in rat cerebral cortex. J Pharmacol Exp Ther 185:371-378

Pollard H, Schwartz J C (1987) Histamine neuronal pathways and their functions. Trends Neurosci 10:86-89

Reinhart P H, Levitan I B (1995) Kinase and phosphatase activities intimately associated with a reconstituted calcium-dependent potassium channel. J Neurosci 15:4572-4579

Robinson J P, Kendall D A (1989) Inositol phospholipid hydrolysis and potentiation of cyclic AMP formation by noradrenaline in rat cerebral cortex slices are not mediated by the same α-adrenoceptor subtypes. J Neurochem 52:690-698

Sah P (1996) Ca^{2+}-activated K^+ currents in neurones: types, physiological roles and modulation. Trends Neurosci 19:150-154

Santoro B, Grant S G N, Bartsch D, et al (1997) Interactive cloning with the SH3 domain of N-src identifies a new brain specific ion channel protein, with homology to eag and cyclic nucleotide-gated channels. Proc Natl Acad Sci USA 94:14815-14820

Santoro B, Liu D T, Yao H, et al (1998) Identification of a gene encoding a hyperpolarization-activated pacemaker channel of brain. Cell 93:717-729

Shacter E, Stadtman E R, Jurgensen S R, et al (1988) Role of cAMP in cyclic cascade regulation. Meth Enzymol 159:3-19

Sim J A, Gerber U, Knöpfel T, et al (1992) Evidence against a role for protein kinase C in the inhibition of the calcium-activated potassium current I_{AHP} by muscarinic stimulants in rat hippocampal neurons. Eur J Neurosci 4:785-791

Steriade M, McCarley R W (1990) Brainstem control of wakefulness and sleep. Plenum Publishing Corp, New York.

Stevens C F (1994) CREB and memory consolidation. Neuron 13:769-770

Stone E A, Herrera A S (1986) Alpha-adrenergic modulation of cyclic AMP formation in rat CNS: highest level in olfactory bulb. Brain Res 384:401-403

Storm J F (1990) Potassium currents in hippocampal pyramidal cells. Prog Brain Res 83:161-187

Storm J F, Winther T, Pedarzani P (1996) H-Current modulation by norepinephrine, serotonin, dopamine, histamine and cyclic-AMP analogues in rat hippocampal neurons. Soc Neurosci Abstr 22:1444

Tang W J, Gilman A G (1991) Type-specific regulation of adenylyl cyclase by G protein beta gamma subunits. Science 254:1500-1503

Verney C, Baulac M, Berger B, et al (1985) Morphological evidence for a dopaminergic terminal field in the hippocampal formation of young and adult rat. Neuroscience 14:1039-1052

Walaas S I, Greengard P (1991) Protein phosphorylation and neuronal function. Pharmacol Rev 43:299-349

Woodward D J, Moises H C, Waterhouse B D, et al (1979) Modulatory actions of norepinephrine in the central nervous system. Fed Proc 38:2109-2116

Zifa E, Fillion G (1992) 5-Hydroxytryptamine receptors. Pharmacol Rev 44:401-458

Three Types of Cerebellar Voltage-Gated K+ Currents Expressed in *Xenopus* Oocytes

N. HOSHI H. TAKAHASHI, S. YOKOYAMA, and H. HIGASHIDA

Department of Biophysical Genetics, Kanazawa University Graduate School of Medicine,13-1 Takara-machi Kanazawa Japan 920-8640

Key words. Voltage-gated K+ current, Xenopus oocyte, $I_{K(SO)}$

Recently, a low-threshold, noninactivating current has been found in cerebellar granule neurons, designated $I_{K(SO)}$ (Watkins and Mathie 1996). To characterize the noninactivating current in the cerebellum, we examined cerebellar K+ currents heterologously expressed in *Xenopus* oocytes by injecting poly(A)+ RNA that was obtained from the cerebellum. Our results of the expression of various sucrose density fractions of the cerebellar poly(A)+ RNA revealed that the distinct RNA species was responsible for each K+ current.

Xenopus oocytes that have been injected with poly(A)+ RNA (5-10 ng) from rat cerebellum elicited outward K+ currents of 1-5 μA on a depolarizing step to +40 mV with an average peak amplitude of 2.5 ± 0.7 μA ($n=7$) from a holding potential of -100 mV or 1.7 ± 0.5 μA ($n=7$) from a holding potential of -60 mV, respectively. In contrast, identical depolarizing steps yielded only a small outward current (30 ± 14.8 nA, $n=14$) in noninjected oocytes.

Further pharmacological and kinetical characterization of the outward K+ currents dissected them into three components. Tetraethylammonium (TEA) partially suppressed the outward K+ currents, at a concentration of 10 mM (Fig. 1C, D). We refer this current as $I_{S/TEAS}$, where S and TEAS represent the sustained and TEA-sensitive current (Fig. 2E). The current remaining in the presence of TEA apparently consists of both transient and sustained components (Fig. 1C). Figure 1D shows the sustained component ($I_{S/TEAR}$), where TEAR represents the TEA-resistant current. The TEA-resistant and transient current ($I_{T/TEAR}$) can be calculated by subtracting between currents elicited from two different holding potentials of -100 and -60 mV (C-D), as shown in Fig. 1F.

TEA applied at 10 mM to the perfusion solution revealed that 70% of the current measured at the end of 500-ms test pulses was TEA sensitive. The dose-response relationship of TEA corresponded to a single binding site with the $K_{1/2}$ value of 0.065 ± 0.03 mM ($n=12$). This result suggests that the $I_{S/TEAS}$ consisted of a homogenous group of TEA sensitive channels which was distinct from the TEA resistant K^+ currents ($I_{S/TEAR}$ and $I_{T/TEAR}$). This result also shows that the voltage-gated K^+ current derived from cerebellar poly(A)$^+$ RNA is composed of less complicated channel species than those derived from brain poly(A)$^+$ RNA (Hoger et al. 1991), which is advantageous to further analyses.

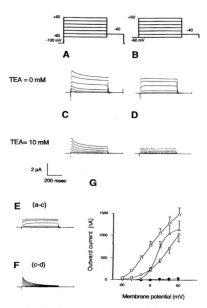

Fig. 1A-G. Voltage-activated membrane currents recorded in *Xenopus* oocytes injected with total cerebellar mRNA. Records of **A-F** were obtained from a single cell. *Uppermost*, voltage protocols for **A** and **C** (*left*) or for **B** and **D** (*right*). Depolarizing steps of 500 msec starting from -60 mV to +60 mV of membrane potentials at increment of 20 mV were applied. Superimposed current traces of the same oocyte were recorded in the absence (**A** and **B**) or presence (**C** and **D**) of 10 mM TEA. Traces in **E** are currents subtracted those in **C** from those in **A**, showing the TEA sensitive component. Traces in **F** are currents subtracted those in **D** from those in **C**, revealing the rapidly inactivating TEA-resistant component. **G** Current-voltage relationships of the three components of currents measured in 7 oocytes injected with mRNA as follows: the TEA-sensitive component (*open triangle*) calculated as in **E**; the TEA resistant inactivating component (*open circle*) as in **F**; and the TEA resistant sustained component (*open square*) as in **D**. The average current amplitude evoked by each voltage step is plotted against voltage. *Closed circles* show currents in non-injected control oocytes (*n*=14). Each point is the mean value±S.E.M.

Next, we addressed the question as to how much the two TEA resistant components differ one from other. To answer this question we examined the sensitivity of the TEA-resistant currents to another K^+ antagonist, 4-aminopyridine (4-AP). The $I_{T/TEAR}$ is reversibly blocked by 4-AP. For its inhibition >0.1 mM was required. The $I_{S/TEAR}$, however, remained unaltered under the same condition. Altogether, the results suggest that the outward current expressed with total cerebellar poly(A)$^+$ RNA in the oocytes is composed of at least three distinct currents.

Sucrose gradient fractionation also supported the concept that there are three groups of channels. The channel activities for $I_{T/TEAR}$ were recovered at about the 4 kb poly(A)$^+$ RNA fraction, $I_{S/TEAS}$ at 6 kb and $I_{S/TEAR}$ at about 8 kb. The fractionated RNA expression allowed us to measure each group of currents with little contamination of other currents. We then compared the voltage-dependent kinetics and TEA sensitivity between three types of currents and those of cloned voltage-gated K$^+$ currents. We found that $I_{S/TEAS}$ was almost indistinguishable from Kv3.1 currents, and $I_{T/TEAR}$ had a voltage dependence that differed from Kv4.2 as reported to be expressed in the cerebellum (Yokoyama et al. 1989; Sheng et al. 1992).

The $I_{S/TEAR}$ did not fall into any of the cloned channel characteristics. Figure 2A shows prepulse inactivation for $I_{S/TEAR}$. The $I_{S/TEAR}$ evoked by depolarization steps to +60 mV after a holding membrane potential, even as much as +20 mV for 18.5 s, did not inactivate appreciably (less than 10% of the control amplitude). The $I_{S/TEAR}$ began to be activated at a more positive voltage than -60 mV and did not reach saturating conductance even at a voltage more positive than +60 mV (Fig. 2B). The activation and inactivation data for $I_{S/TEAR}$ were plotted in Fig. 2C, and the parameters derived from a modified Boltzmann fitting function are $V_{1/2}$ of 3.5±4.1 mV and a slope factor of 24.1±2.7 mV ($n=9$).

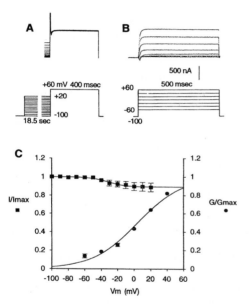

Fig. 2A-C. Voltage dependence of the TEA resistant sustained K$^+$ current. Currents of oocytes injected with 8-kb mRNA fractions after varied prepulses (**A**) or during varied test depolarization (**B**). A Lower voltage pulse protocol used to assess prepulse inactivation. Prepulses (18.5 s) to voltage between -100 mV and +20 mV were followed by a 400-ms test pulse to +60 mV. The holding potential was -100 mV. *Upper:* traces represent currents obtained from an oocyte expressing $I_{S/TEAR}$. Note that linear leakage current is not subtracted. **B** Currents (*upper*) elicited by lower voltage pulse for 500 ms from -60 to +60 mV from a holding potential of -100 mV. **C** Conductance-voltage relations (; $n=10$) and prepulse inactivation properties (*closed square*; $n=3$) of $I_{S/TEAR}$. *closed circle* Each point represents the mean value; error bar shows S.E.M. The normalized conductance to the maximal conductance (G/Gmax) is plotted as a function of the membrane potential. For the prepulse inactivation curve, the leak-subtracted current amplitude of the end of test pulse was normalized to the maximal current after a prepulse to -100 mV. The *solid curves* represent the best-fitting modified Boltzmann functions to the averaged data.

The low-threshold noninactivating K$^+$ currents in cerebellar granule neurons have been reported as $I_{K(SO)}$ (Watkins and Mathie 1996) and are inhibited by muscarinic agonists. We tested whether the stimulation of PI turnover could inhibit $I_{S/TEAR}$. The focal application of 5-HT transiently inhibited this current (Fig. 3).

In conclusion, we have demonstrated that *Xenopus* oocytes injected with cerebellar poly(A)$^+$ RNA exhibit three types of voltage-gated K$^+$ currents: a Kv3.1-like TEA-sensitive delayed rectifier current, a TEA-resistant A-current, and an $I_{K(SO)}$-like noninactivating current. These classified current components will be utilized for further characterization at the molecular level.

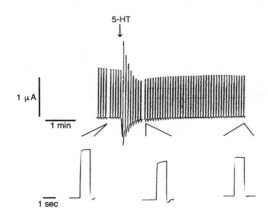

Fig. 3. Effect of 5-HT on the $I_{S/TEAR}$. (*Top*) Current responses to focal application of 5-HT. The currents were activated by 1 sec depolarizing steps to +20 mV from a holding potential of -60 mV; 10 μl of 5-HT solution of 100 μM were applied at the *arrow*. (*Bottom*) Expanded current traces in response to the depolarizing steps recorded at before 5-HT, during the inhibition, and after recovery.

Acknowledgments. This work was supported by grants from the Ministry of Science, Education and Culture of Japan.

References

Hoger JH, Rudy B, Lester HA, et al. (1991). Characterization of maintained voltage-dependent K$^+$-channels induced in *Xenopus* oocytes by rat brain mRNA. Mol Brain Res 10:1-11

Sheng M,Tsaur, M-L, Jan YN, et al (1992) Subcellular segregation of two A-type K$^+$ channel proteins in rat central neurons. Neuron 9:271-284

Watkins CS, Mathie A (1996). A non-inactivating K$^+$ current sensitive to muscarinic receptor activation in rat cultured cerebellar granule neurons. J. Physiol. (Lond.) 491:401-412

Yokoyama S, Imoto K, Kawamura T, et al (1989). Potassium channels from NG108-15 neuroblastoma-glioma hybrid cells: primary structure and functional expression from cDNAs. FEBS Lett. 259:37-42

Facilitatory Effect of Calmodulin-Dependent Protein Kinase on the K+-Current Responses to Dopamine, Acetylcholine, and Phe-Met-Arg-Phe-NH₂ in the Ganglion Cells of *Aplysia*

S. KAWASAKI, S. KIMURA, R. FUJITA, K. TAKASHIMA, K. SASAKI, AND M. SATO

Department of Physiology, School of Medicine, Iwate Medical University, 19-1 Uchimaru Morioka 020-8505, Japan

Key words. Calmodulin-dependent protein kinase, Acetylcholine receptor, Dopamine receptor, Phe-Met-Arg-Phe-NH₂-receptor, G protein, Potassium channel

1 Introduction

Formation of either long-term potentiation (LTP) or long-term depression (LTD) of the synaptic transmission is thought to be the primary requirement for learning and memory. These LTP and LTD are usually produced by the activation of various enzymes at both pre- and postsynaptic loci after the transsynaptic stimulation of the receptors. We previously reported that acetylcholine (ACh)-, dopamine (DA)-, histamine (HA)-, and Phe-Met-Arg-Phe-NH₂ (FMRFamide)-induced K+-current responses are all mediated by common GTP-binding protein G_i or G_o, irrespective of their kinds of transmitters or types of receptors (Sasaki and Sato 1987). Furthermore, we recently reported that all these K+-current responses are markedly depressed by the activation of intracellular protein kinase A or C (Sasaki et al. 1997). At present we are considering the possibility that other kinases, such as calmodulin-dependent protein kinase (CaMK) and protein tyrosine kinase, or various protein phosphatases may also be activated simultaneously after the stimulation of these receptors, producing the primary reactions necessary for the formation of LTP or LTD (Kawasaki et al. 1998). This study examines the role of CaMK activation on the DA-, ACh-, and FMRFamide-induced K+-current responses in the ganglion cells of *Aplysia*, and discusses the possible acting site of this enzyme in the signal-transducing pathway involved in these receptor-induced responses.

2 Role of CaM Kinase Activation

Stimulation of either one of the receptors for DA, ACh, and FMRF amide induced a common K^+-current response in identified neurons of *Aplysia* ganglia under the conventional voltage clamp using two microelectrodes. Intracellular injection of KN-93, a specific inhibitor of CaMK II, significantly depressed all these responses. Intracellular concentration of KN-93 was estimated to be 100 μM. Furthermore, extracellular application of 300 μM W-7, an antagonist against calmodulin, reversibly and dose-dependently depressed the K^+-current responses induced by ACh, DA, or FMRF amide. On the contrary, intracellular injection of exogenous calmodulin significantly augmented the receptor-induced K^+-current responses. These results suggested that intrinsic CaMK II is activated by the stimulation of these receptors, leading to the facilitation of the receptor-induced K^+-current responses.

To investigate the possible acting site of CaMK II, we injected guanosine thiotriphosphate (GTPγS) into the cells that produced the K^+-current responses to DA, ACh, or FMRFamide. GTPγS is known to directly activate the G-protein and produce the irreversible opening of the K^+ channels coupled to the receptors. Surprisingly, application of 300 μM W-7 did not at all depress the GTPγS-induced K^+-current response, even though it markedly depressed the K^+-current responses induced by receptor stimulation. These results suggest that the acting site of CaMK II is not the K^+ channel itself but is somewhere between the receptors and G protein involved in the functional coupling of these molecules.

3 Role of Protein Phosphatase

We further attempted to identify a type of protein phosphatase which would dephosphorylate the protein molecule phosphorylated by CaMK II. Intracellular injection of cypermethrin, an inhibitor of calcineurin (protein phosphatase 2B), significantly augmented the receptor-induced K^+-current responses. Intracellular concentration of cypermethrin was estimated to be 1 μM. Extracellular application of 3 μM FK-506, another inhibitor of calcineurin, similarly augmented the K^+-current responses. In contrast, intracellular injection of Okadaic acid, an inhibitor of protein phosphatase 1 and 2A, did not affect the receptor-induced K^+-current responses. Furthermore, intracellular application of the catalytic subunit of calcineurin markedly depressed the receptor-induced K^+-current responses. On the contrary, it did not at all depress the GTPγS-induced K^+-current response, indicating that calcineurin did not depress the opening of the K^+-channel itself.

4 Discussion

Our results suggest that the receptor-induced K^+-current responses are reciprocally regulated by both CaMK II and calcineurin; the former augments the receptor-induced K^+-current responses while the later depresses them. Wakamori et al. reported that receptor-induced activation of the CAMK II can open a certain K^+ channel in hippocampal CA1 neurons (Wakamori et al. 1993). Usually, most studies using single channel recording have also indicated that phosphorylation of the K^+ channel itself by various protein kinases exerts either augmentation or inhibition of the channel activities depending on the kinds of kinases examined. Thus, many workers place importance mainly on the channel molecule rather than other signaling proteins as a possible acting site of various kinases. However, our present results suggested that CaMK II would act on the coupling site somewhere between the receptors and G protein but not the K^+ channel itself, leading to facilitation of the coupling efficiency between these molecules. Although we have not identified the molecules phosphorylated by CAMK II which are crucial for the facilitation, one of the candidates for this substrate molecule could be the so-called RGS (regulators of G-protein signaling), which is a recently reported protein family regulating GTPase activity of the G protein (Saitoh et al. 1997).

In contrast to the facilitatory action of CaMK II shown in the present study, we previously reported the inhibitory action of protein kinase A or C on the receptor-induced K^+-current responses to DA, ACh, HA, and FMRFamide. These protein kinases A and C also act on a site somewhere between the receptors and G protein, leading to suppression of the coupling efficiency between the receptor and G protein. Protein kinase C is known to be activated by diacylglycerol as a result of phospholipase C activation after the receptor stimulation, while CaMK II is activated by an increase in concentration of Ca^{2+} released from the intracellular Ca^{2+} store as a result of the same phospholipase C activation. Therefore, it should be noted that CaMK II and protein kinase C, respectively, form different feedback pathways which act reciprocally on the coupling efficiency between the receptor and G protein. We speculate that various kinds of kinases including not only protein kinase C and CaMK II but also other kinases are simultaneously activated subsequent to the stimulation of the receptors coupled with the K^+ channel and that the summated effect may result in the regulation of the receptor-induced K^+ current responses in the normal state. Imbalance among these enzyme activities produced by the receptor stimulation may cause the LTP and LTD.

References

Kawasaki S, Sasaki K, Kimura S, et al (1998) Augmenting effect of sequential activation of phospholipase C and CaM-kinase on the receptor-induced K^+-current responses observed in *Aplysia* neurons. Jpn J Physiol 48 supple: s98.

Saito O, Kubo Y, Miyatani Y, et al (1997) RGS8 accelerates G-protein-mediated modulation of K$^+$ currents. Nature 390:525-529

Sasaki K, Sato M (1987) A single GTP-binding protein regulates K$^+$-channels coupled with dopamine, histamine and acetylcholine receptors. Nature 325:259-262

Sasaki K, Kawasaki S, Kimura S, et al (1997) Functional uncoupling between receptor and G-protein as the result of PKC activation, observed in *Aplysia* neuron. Jpn J Physiol 47:241-249

Wakamori M, Hidaka H, Akaike N (1993) Hyperpolarizing muscarinic responses of freshly dissociated hippocampal CA1 neurons. J Physiol 463: 85-604

Introductory Review: Calcium Channels and Modulation

L. GANDÍA[1], A. ALBILLOS[1], C. MONTIEL[1] and A. G. GARCÍA[1,2]

[1]Instituto de Farmacología Teófilo Hernando, Departamento de Farmacología, Facultad de Medicina, Universidad Autónoma de Madrid, Arzobispo Morcillo 4, 28029 Madrid, Spain
[2]Servicio de Farmacología Clínica and Instituto de Gerontología, Hospital Universitario de la Princesa, Diego de León 62, 28006 Madrid, Spain

1 Introduction

Voltage-dependent Ca^{2+} channels are strategically located in the plasmalemma of excitable cells to initiate, mediate, or regulate important and different Ca^{2+}-dependent functions, i.e., cell excitability, muscle contraction, neurotransmitter and hormone release, or gene transcription. The combination of patch-clamp techniques (Hamill et al. 1981) with various marine and insect ω-toxins (Olivera et al. 1994) as well as molecular biology approaches (Striessnig et al. 1998), have revealed a considerable diversity in their primary structures, biophysics, pharmacology, regulation, and expression in different mammalian tissues and species. This introductory review to the section of "Ca^{2+} currents and modulation" of this book deals with aspects related to the diversity of Ca^{2+} channels, their kinetic and pharmacological characteristics, their species differences, their molecular structure and their regulation by voltage, $[Ca^{2+}]_i$, phosphorylation and intracellular second messengers. Their regulation by neurotransmitters via a membrane-delimited G-protein pathway is discussed in a more detailed manner in the chapter "Exocytosis calcium channels: autocrine/paracrine modulation" in this section of the book. Several reviews that deal in more detail with these different aspects of Ca^{2+} channels are available (Hagiwara and Byerly 1981; Carbone and Swandulla 1989; Scott et al. 1991; Tsien et al. 1991; Olivera et al. 1994; Hille, 1994; Hoffmann et al. 1994; Wickman and Clapham 1995; García et al. 1997; Uchitel, 1997; Dolphin 1998).

Table 1.- Biophysical, kinetics and pharmacology of the various subtypes of voltage-dependent calcium channels

Channel subtype	Activation	Conductance	Inactivation kinetics	Pharmacology
Low threshold (LVA)				
T	>-60mV	6-9 pS	τ=5-10 ms Voltage-dependent	Ni²⁺ Amiloride Mibefradil
High threshold (HVA)				
L	>-50mV	18-28 pS	τ>200ms Calcium-dependent	Co²⁺, Mn²⁺, Cd²⁺ Dihydropyridines Verapamil Diltiazem
N	>-50mV	7-14 pS	τ=50ms Calcium-dependent	Co²⁺, Mn²⁺, Cd²⁺ ω-conotoxin GVIA ω-conotoxin MVIIA ω-conotoxin MVIIC ω-conotoxin MVIID
P	>-45mV	9,14,19 pS	τ=1 s	Co²⁺, Mn²⁺, Cd²⁺ FTX; sFTX ω-agatoxin IVA (nM) ω-conotoxin MVIIC
Q	—	—	—	ω-conotoxin MVIIC ω-conotoxin MVIID ω-agatoxin IVA (µM)
R?	—	—	—	Ni²⁺>Cd²⁺

2 Biophysics and Pharmacology

Up to now, the combination of patch-clamp techniques and pharmacological probes have led to the identification and characterization of at least six subtypes of voltage-dependent Ca^{2+} channels, which have been termed L, N, P, Q, R, and T. According to their range of activation these channels are classified in two groups, low-voltage-activated (LVA) and high voltage-activated (HVA) Ca^{2+} channels. Table 1 summarizes the main biophysical and pharmacological characteristics of the Ca^{2+} channels identified up to now. A single LVA Ca^{2+} channel has been identified (Carbone and Lux 1984); it was termed T (for "Transient" or "Tiny"). These channels have a low threshold for activation and a fast inactivation, thus generating a transient current; inactivation occurs at hyperpolarizing potentials (-60 to -50 mV). Their single-channel conductance is around 8 pS. They have a similar permeability for Ca^{2+} and Ba^{2+}, and they are blocked better by Ni^{2+} than by Cd^{2+} (Fox et al. 1987). They are also blocked by octanol and amiloride. Recently, a nonpeptide molecule, mibefradil, has been characterized as a selective T-type Ca^{2+} channel blocker (Mishra and Hermsmeyer 1994); however, mibefradil also blocks HVA channels (Bezprozvanny and Tsien 1995)

In contrast to LVA channels, HVA channels (L, N, P, Q, R) activate by strong depolarizing steps, have higher permeability to Ba^{2+} than to Ca^{2+}, and have a higher sensitivity to Cd^{2+} than to Ni^{2+} (Fox et al. 1987). The major differences between these subtypes are related to their inactivation kinetics and their pharmacological properties. L-type (for "Long Lasting") Ca^{2+} channels show little inactivation (τ_{inact}>500 ms) and have low sensitivity to depolarised holding potentials. Single-channel conductance is around 18-25 pS. They are present in excitable and nonexcitable cells and constitute the main pathway for Ca^{2+} entry in heart and smooth muscle; they also serve to control hormone and transmitter release from endocrine cells and some neuronal preparations. They are highly sensitive to 1,4-dihydropyridines (DHPs), both agonists (i.e. Bay K 8644, FPL64176) and antagonists (i.e. nifedipine, nimodipine, furnidipine). Other organic compounds such as arylalkylamines (i.e. verapamil) and benzothiazepines (i.e. diltiazem) are particularly potent in cardiac muscle cells, where they exert negative inotropic effects (Fleckenstein 1983). Some piperazine derivatives (cinnarizine, flunarizine, dotarizine, R56865, lubeluzole) block L-channels; however, they also block other subtypes of Ca^{2+} channels and thus have been named "wide-spectrum" Ca^{2+} channel blockers (Gárcez do Carmo et al. 1994; Villarroya et al. 1995, 1997).

N-type Ca^{2+} channels (for "neuronal" or "Non-L-Non-T") are distinguishable from L-type for their faster inactivation kinetics (τ_{inact} 50-80 ms). Single-channel conductance of N channels is about 13 pS. The *Conus geographus* toxin ω-conotoxin GVIA blocks them irreversibly (Nowycky et al. 1985; Kasai et al. 1987; Olivera et al. 1994); the Conus magus ω-conotoxin MVIIA blocks them reversibly (Valentino et al. 1993; Vega et al. 1995). Other wide-spectrum toxins isolated from the venom of Conus magus snails as ω-conotoxin MVIIC and ω-conotoxin MVIID (Hillyard et al. 1992; Monje et al. 1993) can also block N-channels in a nonselective manner. ω-Conotoxin MVIID blocks the N-channel of bovine chromaffin cells in a readily reversible way, while blockade of this channel by ω-conotoxin MVIIC is more stable (Gandía et al. 1997).

P-type Ca^{2+} channels were first described by Llinás et al. (1989) in cerebellar Purkinje cells, in which Ca^{2+} currents were resistant to blockade by DHPs and ω-conotoxin GVIA. The toxin fraction from the venom of the funnel web spider *Agelenopsis aperta* (FTX) was found to effectively block this resistant current. These results led these authors to suggest the existence of a new subtype of HVA Ca^{2+} channel, which was termed P (for "Purkinje"). P channels are relatively insensitive to changes in the holding potential, and do not inactivate during depolarizing steps (Regan 1991; Mintz et al. 1992). Multiple single channel conductances have been described for P-type Ca^{2+} channels (Usowicz et al. 1992; Umemiya and Berger, 1995). Although P-type Ca^{2+} channels were first described by their blockade by FTX and its synthetic analog sFTX, these two toxins were found to block not only P channels but also other voltage- and ligand-activated whole-cell currents (Scott et al. 1992). These results led to a further purification of the venom and three types of toxins were found (acylpolyamines, ω-agatoxins and μ-conotoxins). The ω-agatoxins (Adams et al. 1990) have been shown to specifically block different subtypes of Ca^{2+} channels (see above). The fraction identified as ω-agatoxin IVA selectively blocks P-type Ca^{2+} channels, with a K_D of 1-2 nM (Mintz et al. 1992; Mintz and Bean, 1993), and thus in the nanomolar range (<30-100 nM) this toxin is actually accepted to be the probe to identify the presence of P-type Ca^{2+} channels. At higher concentrations (>100 nM), ω-agatoxin IVA blocks not only P channels but also Q-type Ca^{2+} channels (see below). P-channels can also be blocked in a nonselective manner by the Conus magus snail toxins with wider spectrum of action, ω-conotoxin MVIIC (Hillyard et al. 1992; Monje et al. 1993) and ω-conotoxin MVIID (Gandía et al. 1997).

In many neuronal preparations, a significant component of the whole-cell current through Ca^{2+} channels is resistant to blockade with DHPs, ω-conotoxin GVIA and ω-agatoxin IVA (<100 nM), suggesting the presence of a subtype of Ca^{2+} channel different from L-, N-, and P-types. The isolation, purification and synthesis of the toxin from the marine snail *Conus magus* ω-conotoxin MVIIC (Hillyard et al. 1992; Monje et al. 1993) led to the identification and characterization of a new subtype of HVA channel termed Q (Randall et al. 1993; Wheeler et al. 1994; Randall and Tsien 1995). Characterization of Q-type Ca^{2+} channels is mostly based on pharmacological criteria. As described, Q-type channels are resistant to blockade by DHPs, ω-conotoxin GVIA, and low doses (<100 nM) of ω-agatoxin IVA, but are sensitive to ω-conotoxin MVIIC (0.3-3 μM). Increasing concentrations of ω-agatoxin IVA (up to 2 μM) can also block Q-channels (Wheeler et al. 1994). It should be noted that these toxins used to identify Q-channels are not selective for this subtype of channels and they also block, in a nonselective manner, N- and P-type channels. Other toxins have also been described as Q-type Ca^{2+} channel blockers, including the *Conus magus* snail toxin ω-conotoxin MVIID (Monje et al. 1993; Gandía et al. 1997).

In neuronal tissues, a residual Ca^{2+} current characterized by its insensitivity to blockade by DHPs, ω-conotoxin GVIA, ω-agatoxin IVA, and ω-conotoxin MVIIC has also been described and termed "R-type" (for "Resistant"; Randall et al. 1993; Randall and Tsien 1995). This new subtype of Ca^{2+} channel belongs to the HVA group, is rapidly inactivating (τ=22 ms), and is more sensitive to blockade by Ni^{2+} (IC_{50}=66 μM) than by Cd^{2+}. Because the high concentrations of Ba^{2+} used to study Ca^{2+} channel currents prevent the binding of ω-conotoxin MVIIC to its receptor, we made a call of caution to the fact that toxin-resistant currents might be an artefact of experimental protocols (Albillos et al. 1996).

3 Some Curious Differences Among Species

Drastic species differences in the subtypes of Ca^{2+} channels expressed by different cell types have been found. Detailed comparative electrophysiological studies among six mammalian species have been performed only in adrenal medullary chromaffin cells (see Fig. 1 in chapter from Garcia et al.). L-type Ca^{2+} channels account for nearly half of the whole-cell Ca^{2+} channel current in the cat (Albillos et al. 1994), rat (Gandía et al. 1995), and mouse chromaffin cells (Hernández-Guijo et al. 1998). In pig (Kitamura et al. 1997), bovine (Gandía et al. 1993; Albillos et al. 1993), and human species (Gandía et al. 1998), L-channels carry only 15-20% of the whole-cell Ca^{2+} current.

The N-type Ca^{2+} channels also show a high interspecies variability. In the pig it carries as much as 80% of the whole-cell Ca^{2+} channel current (Kitamura et al. 1997) and in the cat 45% (Albillos et al. 1994); in bovine (López et al. 1994b), rat (Gandía et al. 1995), mouse (Hernández-Guijo et al. 1998), and human chromaffin cells (Gandía et al. 1998) the N-type fraction accounts for 30% of the whole-cell Ca^{2+} channel current.

P-type Ca^{2+} channels have proven difficult to characterize in chromaffin cells. Through the use of the synthetic funnel-web toxin sFTX (Gandía et al. 1993), 1 μM ω-agatoxin IVA (Albillos et al. 1993), or 100 nM ω-agatoxin IVA (Artalejo et al. 1994) as much as 40-55% of the whole-cell Ca^{2+} channel current was initially attributed to P channels. Later on we learned that concentrations of ω-agatoxin IVA greater than 10-20 nM in addition of P channels (Mintz et al. 1992), also block Q-channels (Wheeler et al. 1994). Thus, nanomolar concentrations of ω-agatoxin IVA known to block fully and selectively P channels cause only 5-10% blockade of Ca^{2+} channel current in bovine chromaffin cells (Albillos et al. 1996). In cat chromaffin cells, combined ω-conotoxin GVIA plus nisoldipine blocked 90% of the current, leaving little room for P channels (Albillos et al. 1994). In rat (Gandía et al. 1995) and mouse (Hernández-Guijo et al. 1998), the ω-agatoxin IVA-sensitive current fraction was only 10-15%. Thus, in all species studied it seems that P channels are barely expressed, if at all, in their chromaffin cells. This, together with the difficulty of separating the α_{1A} subunit into P- and Q-channels (Sather et al. 1993), suggests the convenience of speaking of P/Q channels rather than of two separate Ca^{2+} channel subtypes.

The P/Q-type Ca^{2+} channel is pharmacologically isolated by 2 μM ω-conotoxin MVIIC or ω-conotoxin MVIID, or by 2 μM ω-agatoxin IVA. In bovine chromaffin cells, ω-conotoxin MVIID blocks the N current reversibly while ω-conotoxin MVIIC does so irreversibly (Gandía et al. 1997). Thus, the use of ω-conotoxin MVIID followed by its washout can be a convenient tool to isolate the P/Q channel. The blocking effects of ω-conotoxin MVIIC are extraordinarily slowed and decreased in the presence of excessive concentrations (i.e. more than 2 mM) of Ba^{2+} (Albillos et al. 1996; McDonough et al. 1996) or Ca^{2+} (Vega et al. 1995). Taking into consideration these methodological problems, we believe that the fraction of current carried by P/Q channels amounts to 50% in bovine chromaffin cells (Albillos et al. 1996). This fraction is even higher (60%) in human chromaffin cells (Gandía et al. 1998). The opposite occurs in pig (Kitamura et al. 1997) and cat chromaffin cells (Albillos et al. 1994), where P/Q channels carry only 5% to the current. Finally, in rat chromaffin cells P/Q channels contribute 20% to the current (Gandía et al. 1995), and in the mouse 30% (Hernández-Guijo et al. 1998).

4 Molecular Structure

Voltage-gated Ca^{2+} channels are basically a multiple subunit protein complex consisting of a pore-forming α_1 subunit and several auxiliary subunits, which include the intracellular β subunit and a disulfide-linked α_2/δ subunit (Hoffmann et al. 1994). In some tissues, a fifth subunit may also form part of the channel complex, such as the transmembrane γ subunit found in skeletal muscle or the neuronal p95 subunit. The functional diversity found among different subtypes of Ca^{2+} channels can be explained according to molecular differences of the channels due to (1) the existence of multiple genes encoding different classes of α_1 and β subunits as well as diverse variants from a single gene generated by alternative splicing, and (2) multiple possible combinations among the subunits that make up the channel complex. Expression studies have shown that voltage-activated channel function, which is typical for the L-, N-, P-, Q-, R-, and T-type Ca^{2+}

Table 2. Voltage-activated Ca^{2+} channel α_1 subunit genes and their functional correlates

Gene product	Functional channel	Tissue
LVA		
α_{1G}	T-type	Brain, Heart
α_{1H}	T-type	Heart
HVA		
α_{1D}	L-type	Brain, pancreas, PC12 and GH3 cells
α_{1C}	L-type	Heart, brain, aorta, lung, fibroblast, kidney, PC12 and GH3 cells
α_{1S}	L-type	Skeletal muscle
α_{1F}	L-type	Retina
α_{1B}	N-type	Brain, peripheral neurons and PC12 cells
α_{1A}	P/Q-family	Brain, cerebellum, Purkinje and granule cells, kidney, PC12 cells
α_{1E}	R-type ?	Brain, heart

HVA, high voltage-activated Ca^{2+} channel; LVA, low voltage-activated Ca^{2+} channel.

channels, is carried by the corresponding α_1 subunit. This subunit confers the characteristic pharmacological and functional properties of Ca^{2+} channels, although their function and properties are modulated by association with the other auxiliary subunits.

The Ca^{2+} channel α_1 subunits isolated, sequenced and expressed so far can be grouped in nine different classes (Table 2). It is generally agreed that some of these subunits encode functional L (α_{1S}, α_{1C}, α_{1D}, α_{1F}), N (α_{1B}), and P/Q (α_{1A}) - type Ca^{2+} channels. However, the correspondence of α_{1E} to a functional channel remains controversial since it has been suggested that it would encode either one R subtype Ca^{2+} channel or a T subtype. At the beginning of 1998, Pérez-Reyes and colleagues have reported the cloning and expression of a new α_1 subunit referred to as α_{1G}, being this the first member of the low-voltage-activated Ca^{2+} channel family, which clearly appears to support the classical T-type Ca^{2+} channel activity (Pérez-Reyes et al. 1998). The finding of this subunit does not mean that only one subtype of T-type Ca^{2+} channel exists, since the same group has reported a close relative gene highly expressed in heart, referred to as α_{1H}, which also supports a typical T-type channel activity. Table 2 summarizes the nomenclature for the Ca^{2+} channel α_1 genes known to date and relates each gene with their functional correlate Ca^{2+} channel type as well as the major sites of expression. The diversity of α_1 genes found, together with the alternative splicing from each single gene, add a large structural diversity to the multitude of Ca^{2+} channel α_1 gene subproducts.

α_1 subunit

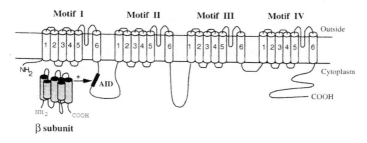

Fig.1. Proposed transmembrane folding model for the Ca^{2+} channel α_1 subunit which contains the α_1 subunit interaction domain (AID) for the β subunit binding.

In spite of the divergence among different α_1 genes, the deduced amino acid sequences of all Ca^{2+} channel α_1 subunits show a basic generalized secondary structure (Fig. 1) that consists of four internal repeated motifs (I-IV). Each motif comprises six α-helical transmembrane segments (S1-S6) including one (S4) that is the positively charged voltage sensor. Also, a specific region exists in all cloned α_1 subunits, with the exception of α_{1G}, which is a highly conserved sequence of the cytoplasmic linker loop between motif I and II. This region is known as the α_1 subunit interaction domain (AID) for the binding of β subunits in the Ca^{2+} channel.

The intracellular β subunit is also represented by a multigene family and, up to now, a minimum of four genes have been established with alternative splicing (Hofmann et al. 1994). The α_2/δ protein is a glycosilated protein derived from a proteolytic cleavage of a single gene with alternative splicing that gives rise to the existence of five messenger RNA species (α_{2a-e}) (Brust et al. 1993). The α_2 and δ polypeptides are connected by a disulfide bond and recent data (Wiser et al. 1996) suggest that the α_2 was entirely located outside the cell while the δ polypeptide anchors the whole protein to the membrane through a single transmembrane domain.

Although all classes of α_1 subunits direct the expression of functional Ca^{2+} channels in different expression systems (i.e., *Xenopus* oocytes, mouse fibroblast L cells or HEK293 cells), the coexpression of several classes of β subunits with different α_1 subunits alters the voltage-dependence, kinetics, and magnitude of the Ca^{2+} channel current, with different effects depending on the particular β and α_1 subunits used. The most complete studies about the role of auxiliary subunits within the Ca^{2+} channel complex have been done with the neuronal α_{1A}, the skeletal muscle α_{1S}, and the cardiac α_{1C} subunits. Thus, the coexpression of four different β subunit genes (β_{1b}, β_{2a}, β_3, and β_4) with the α_{1A} subunit, drastically increases the current amplitude expressed in oocytes. All β subunits trigger significant changes in the voltage-dependence of both activation and inactivation. The α_2/δ_b subunit does not modify the properties of the α_{1A} subunit in the absence of β subunits. However, it increases the β-induced stimulation in current amplitude and regulates the β-induced change in inactivation kinetics. Both α_2/δ_b and β_{1b} slightly modify the sensitivity of the α_{1A} subunit to ω-conotoxin MVIIC (de Waard and Campbell 1995). The mechanisms responsible for the α_2/δ effects are not known, although it has been suggested that α_2/δ subunits may stabilize the incorporation of the channel complexes in the membrane and also increase the binding affinity of channels for ω-conotoxins.

It has also been demonstrated that β subunits increase the peak current induced by the expression of cardiac α_{1C} and that this effect can be correlated with an increase in the number of DHP-binding sites (Pérez-Reyes et al. 1992; Castellano et al. 1993). Because DHPs bind on the α_{1C} subunit, the increase both in the current and in the number of DHP binding sites would suggest that β subunits act by increasing the number of channels present at the plasma membrane. However, gating charge measurements (Neely et al. 1993) and immunoblot analysis (Nishimura et al. 1993) demonstrate that the number of α_{1C} subunits expressed at the plasma membrane was not altered by the β subunits. To reconcile these apparently contradictory results it was hypothesized that β subunits could induce important conformational changes on the α_{1C} subunit, which in turn would increase not only the opening probability of the channel, but also the accessibility of the drug to its binding site.

5 Modulation of Ca²⁺ Channels

Ca^{2+} channels can inactivate in a voltage- or Ca^{2+}-dependent manner. Several neurotransmitters have been shown to influence Ca^{2+} channel currents in either their kinetics or their size. The phosphorylation of Ca^{2+} channels or their coupling with different G-proteins are possibly the two main transduction pathways regulating the kinetics of different subtypes of Ca^{2+} channels. We will briefly refer to these aspects, which have been possible to elucidate because individual Ca^{2+} channel subtypes could be pharmacologically isolated by using the ω-toxins.

5.1 Modulation by Ca²⁺ and Voltage

Ca^{2+} currents inactivate, although at rates considerably slower than Na^+ currents. There is a great range of time constants for inactivation, from 1 ms to 1 s, depending on the cell type. It was partially proposed that Ca^{2+} current inactivation is entirely due to accumulation of Ca^{2+} at the inner surface of the membrane (Brehm and Eckert 1978; Tillotson 1979). This is supported by the fact that Ca^{2+} channels experience almost no inactivation at large positive potentials near the equilibrium potential for Ca^{2+} (E_{Ca}), at which almost no Ca^{2+} enters the cell (Katz and Miledi, 1971). The inward current decays more rapidly using Ca^{2+} instead of Ba^{2+} as charge carrier in *Paramecium* and *Aplysia* neurones; this suggested to Tillotson, Brehm and Eckert that the slower rate of inactivation of Ba^{2+} currents is because Ba^{2+} is not as effective as Ca^{2+} ions in blocking the Ca^{2+} channel. The introduction of buffers to chelate Ca^{2+} slows down the inactivation, again suggesting a role for the $[Ca^{2+}]_i$ in the inactivation process (Gutnick et al. 1989). Conversely, the rapid photorelease of cytosolic Ca^{2+} in DRG cells loaded with the caged Ca^{2+} compound DM-nitrophen led to a rapid inactivation of HVA Ca^{2+} channels (Morad et al. 1988)

The Ca^{2+} sensitivity of HVA channel inactivation may vary in different neurons (Yatani et al. 1983; Akaike et al. 1988; Gutnick et al. 1989), cardiac cells (Campbell et al. 1988), and secretory cells (Satin and Cook, 1989), where Ca^{2+} channel inactivation has been reported to be Ca^{2+}- and voltage dependent. Voltage-dependent inactivation has been visualized by comparing the currents activated with two sequential voltage pulses. If the interval between the pulses is short and does not allow for substantial recovery of inactivated channels, inactivation is evidenced as a decrease of current amplitude during the second activating pulse (Gutnik et al. 1989). The most convincing evidence for the voltage dependence of inactivation, however, comes from the finding that the recovery of inactivated channels (repriming) is sensitive to brief hyperpolarizations to -100 mV (Yatani et al. 1983; Gutnik et al. 1989). Thus, inactivation of HVA channels is not simply due to a buildup of the internal $[Ca^{2+}]$ during channel activity but depends on Ca^{2+} and voltage in a rather complex manner.

Voltage inactivation differs for the different HVA Ca channels. Thus, Ca^{2+} channel currents recorded in bovine chromaffin cells with highly buffered cytosolic Ca^{2+} exhibited steady-state inactivation that followed two components (Bossu et al. 1991). One was half-inactivated at low voltages around -55 mV, affected mainly the initial transient component, and was sensitive to ω-conotoxin GVIA (N-type Ca^{2+} channel). The other affected mainly the sustained component of the Ca^{2+} current, inactivated at voltages around -10 mV, and was sensitive do DHPs (L-type channel). However, in another study, the N-component of the whole-cell Ca^{2+} channel current of bovine chromaffin cells suffered no voltage inactivation (Artalejo et al. 1992).

5.2 Phosphorylation by Protein Kinase A

Noradrenaline was the first neurotransmitter reported to enhance L-type Ca^{2+} current (I_{Ca}) in cardiac myocytes. This effect was exerted via activation of a β-adrenoceptor (Cachelin et al. 1983; Bean et al. 1984) that activates adenylyl cyclase through the GTP-binding protein G_s The increase in cAMP causes activation of protein kinase A and phosphorylation of components of the Ca^{2+} channel.

Histamine and glucagon are also adenylate cyclase-stimulating agents in cardiac cells. In contrast, acetylcholine, adenosine and atrial natriuretic factor decrease the level of cytosolic cAMP. These agents are thought to affect I_{Ca} via the same intracellular cascade as proposed for β-adrenergic stimulation. In cardiac preparations, histamine enhances the upstroke velocity of slow action potentials, elevates the plateau, and increases I_{Ca} (Eckel et al. 1982). The effect of histamine on I_{Ca} is not additive to that of β-adrenergic agonists (Hescheler et al. 1987). In mammalian ventricular preparations, acetylcholine and adenosine decreased I_{Ca} when the current was enhanced by stimulation of the adenylate cyclase (Belardinelli and Isenberg, 1983; Hescheler et al. 1986).

There is less evidence for significant regulation of neuronal voltage-dependent Ca^{2+} channels by changes in intracellular cAMP. Several reports show that neuronal Ca^{2+} currents exhibit little or no sensitivity to cAMP, despite the presence of a DHP-sensitive component of the macroscopic current (McFadzean et al. 1989; Wanke et al. 1994). However, noradrenaline and β-adrenoceptor agonists increase Ca^{2+} currents in hippocampal neurons by a mechanism involving protein kinase A (Gray and Johnson 1987). In addition, the α_1-like subunit of a ω-conotoxin GVIA-sensitive brain Ca^{2+} channel may be phosphorylated by cAMP-dependent protein kinase (Ahlijanian et al. 1991). Expression studies have shown that cAMP increases Ca^{2+} currents, which have the properties of P-type currents expressed in oocytes after injection of cerebellar mRNA (Fournier et al. 1993).

5.3 Phosphorylation by Protein Kinase C

Phorbol esters increase Ca^{2+} channel activity in cardiac cells and sympathetic neurons (Lacerda et al. 1988; Lipscombe et al. 1988), as well as in RINm5F cells (Platano et al. 1996). In addition to this, L-type cardiac Ca^{2+} channels expressed in *Xenopus* oocytes are both enhanced and decreased by PKC activators (Bourinet et al. 1992; Singer et al. 1991). There is also evidence that phorbol esters induce the appearance of new Ca^{2+} channels in *Aplysia* neurons (Strong et al. 1987).

Protein kinase C has been reported to reduce the inhibitory effects of G-proteins on Ca^{2+} channels, possibly by disrupting the coupling of G-proteins to the channels (Swartz, 1993). In hippocampal CA3 and cortical pyramidal neurons, activation of protein kinase C enhances current through N-type Ca^{2+} channels and, in addition, dramatically reduces the G-protein-dependent inhibition of these same channels by the metabotropic glutamate receptor. In fast excitatory transmission at corticostriatal synapses, protein kinase C activators were also found to reduce the inhibitory effect produced by stimulation of the metabotropic glutamate receptor (Swartz et al. 1993). However, in frog sympathetic neurons, protein kinase C enhances both N- and L- currents (Yang and Tsien 1993). In contrast, the inhibition of Ca^{2+} channels by noradrenaline in sensory neurons is blocked by a specific protein kinase C inhibitor (Rane et al. 1987). This inhibitory modulation involves N-type Ca^{2+} channels (Cox and Dunlap 1992).

5.4 Other Kinases and Second Messengers

Other second messengers and kinases are also involved in the regulation of Ca^{2+} currents in neurons. Nitric oxide has been shown to modulate Ca^{2+} currents in superior cervical ganglion neurons (Chen and Schofield, 1993). Low oxygen tension inhibits L-type Ca^{2+} channels in arterial myocytes by a voltage-dependent mechanism (Franco-Obregón et al. 1995). Some diffusible second messengers different from intracellular Ca^{2+}, cGMP, cAMP or PKC is mediating the oxotremorine-induced inhibition of N- and L-type Ca^{2+} channels in rat sympathetic neurons (Mathie et al. 1992).

Several actions have been reported for the cyclic GMP-dependent protein kinase. It might phosphorylate the class C Ca^{2+} channel α_1-subunit (Hell et al. 1993) on a site different from the C-terminal tail and modulate the inhibition induced by somatostatin on neuronal Ca^{2+} channels (Merinery et al. 1994).

5.5 Dephosphorylation of Ca^{2+} Channels

Okadaic acid, which inhibits phosphatases 1 and 2A, and a specific peptide inhibitor of phosphatase 1, both enhanced Ca^{2+} currents in cerebellar granule neurons (Leighton et al. 1994). The peptide inhibitor of phosphatase 1 also enhanced Ca^{2+} currents in sensory neurons (Dolphin 1991). The Ca^{2+}-activated phosphatase 2B (calcineurin) is likely to mediate some aspects of Ca^{2+}-dependent current inhibition (Armstrong 1989; Kostyuk and Lukyanetz 1993).

5.6 Modulation by neurotransmitters of HVA Ca^{2+} channels

The inhibition by several neurotransmitters of HVA Ca^{2+} channels consists in a membrane-delimited mechanism coupled to G proteins. This modulation is the object of two separate chapters in this book and thus we will not describe it further (see chapters "Exocytosis calcium channels: autocrine and paracrine modulation" and "Synaptic modulation mediated by G-protein-coupled presynaptic receptors").

6 Perspectives

In 12 years of ω-toxins use, at least six subtypes of HVA Ca^{2+} channels have been identified and characterized. New toxins are needed to target selectively the Q-type Ca^{2+} channel without affecting the N- or P type. The R- or T-type channels also need new toxins to characterise their functions. Whether the P and Q channels are the same or separate entities in various cell types remains to be clarified. The question of how many Ca^{2+} channel subtypes remain to be discovered is also relevant. In addition, differences among tissues and cell types for a given Ca^{2+} channel are emerging; L-type Ca^{2+} channels differ from skeletal, to cardiac, to smooth muscles and the brain. Are the Q channels from hippocampal and chromaffin cells identical? Judging from the results of binding experiments it seems that Q-type channels of bovine chromaffin cells may differ from brain Q-type channels (Gandía et al. 1997); the question therefore arises as whether central and peripheral Ca^{2+} channels are similar or whether subtypes of Q-type channels exist.

Another question relates to the observation that different Ca^{2+} channels are required to control exocytosis of the same transmitter (i.e., acetylcholine, catecholamines) in the same cell type, and vary with different tissues and animal species. For instance, the K^+-evoked Ca^{2+} entry in brain cortex synaptosomes is controlled by N channels in the chick and by P-channels in the rat (Bowman et al. 1993). On the other hand, neurotransmitter release at the muscle end plate is controlled by N-channels in fishes (Ahmad and Miljanich, 1988; Fariñas et al. 1992; Sierra et al. 1995) and amphibians (Jahromi et al. 1992) and by P channels in mammals (Wessler et al. 1990). The elucidation of the physiological significance of the drastic differences found in the expression of various Ca^{2+} subtypes in adrenal chromaffin cell of six mammal species constitutes a most interesting challenge. The fact that human chromaffin cell express mostly P/Q channels and pig chromaffin cells mostly N channels surely has physiological relevance for the fine control of the differential exocytotic release of adrenaline or noradrenaline in response to stressful conflicts. A similar puzzling question is offered by the cat chromaffin cell, which expresses about 50% N channels and 50% L channels, and yet the secretory response is controlled by L channels (López et al. 1994a)

Another important question relates to the expression of various channel subtypes in the same secretory cell. Why does exocytosis require Ca^{2+} from different pathways? Is it a safety valve to secure the efficiency of the process? If the N-channel is a part of the secretory machinery, what about the L-, P-, or Q-channels? How close are they from exocytotic active sites? And, most interesting, are the channels of a paraneuronal cell such as the chromaffin cell equally organized as those of brain synapses? Why is the release of noradrenaline controlled by N-channels in sympathetic neurones and by L- or Q-channels in chromaffin cells? In addition , why L-channels dominate the control of the release of noradrenaline and Q-channels that of adrenaline in bovine chromaffin cells? (Lomax et al. 1997). Do action potentials recruit different Ca^{2+} channel subtypes in those two catecholaminergic cell types? Will a K^+ depolarizing stimulus recruit Ca^{2+} channels different from those recruited by action potentials in neurones, or by acetylcholine receptors in chromaffin cells? Is the electrical pattern of different excitable cells causing different secretion patterns by simply recruiting specific Ca^{2+} channels with particular gating and kinetic properties?

Another critical question relates to the development of pharmacology for neuronal Ca^{2+} channels. While L-type Ca^{2+} channels have a rich pharmacology that has provided novel therapeutic approaches to treat cardiovascular diseases, nonpeptide molecules which block or inactivate the N-, P-, Q-, T- or R-channels are lacking. Thus, a major goal for research in this field is the search for selective blockers of specific Ca^{2+} channel subtypes. The recent introduction of mibefradil as a T-type Ca^{2+} channel blocker opens new possibilities to study the functions of these channels. The knowledge of the three-dimensional structure in solution of the different toxins is very important for studying the specificity of their interactions with Ca^{2+} channel subtypes and to define active sites that can serve as models to design and synthesize nonpeptide blockers. The ω-conotoxins are small peptides containing 24-29 amino acid residues. It is interesting that the amino acid sequence of ω-conotoxin MVIIA is much more similar to that of ω-conotoxin MVIIC than to ω-conotoxin GVIA, yet the pharmacology of ω-conotoxin MVIIA is much closer to that of ω-conotoxin GVIA (blockade of N-type channels). Thus, it will be very important to define structural differences determining the toxin selectivity for N- or Q-type Ca^{2+} channels. The three-dimensional structures of ω-conotoxin GVIA (Sevilla et al. 1993; Davis et al. 1993; Pallaghy et al. 1993; Skalicky et al. 1993), ω-conotoxin MVIIA (Kohno et al. 1995), and ω-conotoxin MVIIC (Nemoto et al. 1995; Farr-Jones et al. 1995) have been elucidated. Elucidation of the structures of ω-conotoxin MVIID, ω-agatoxin IVA, and other new toxins will facilitate their comparisons and the definition of structural determinants for specific binding to Ca^{2+} channel subtypes.

Nonpeptide blockers for neuronal Ca^{2+} channels are emerging, but they lack selectivity. For instance, the piperazine derivatives flunarizine, R56865, lubeluzole, and dotarizine are "wide-spectrum" Ca^{2+} channel blockers (Gárcez do Carmo et al. 1993; Villarroya et al. 1995, 1997; Hernández-Guijo et al. 1997). Fluspirilene, a member of the diphenylbutylpiperidine class of neuroleptic drugs (which also includes pimozide, clopimozide and penfluridol) has anti-schizophrenic actions and blocks N-type Ca^{2+} channels in PC12 cells (Grantham et al. 1994). It may be that its neuroleptic properties are due, at least in part, to an inhibition of neuronal N-type Ca^{2+} channels. Thus, inhibition (or facilitation) of specific neurotransmitter release by selective blockers (or activators) of Ca^{2+} channels may have functional and therapeutic consequences. For instance, synthetic ω-conotoxin MVIIA protected hippocampal CA1 pyramidal neurons from damage caused by transient, global forebrain ischemia in the rat (Valentino et al. 1993). Selective blockade of N-type Ca^{2+} channels may also be beneficial in treatment of specific pain syndromes (Olivera et al. 1994). Thus, intrathecal administration of as little as 0.3 μg of ω-conotoxin MVIIA completely suppressed the nociceptive responses in the rat hindpaw formalin test. Tactile allodynia was also selectively abolished in a rat neuropathic pain model by intrathecal administration of ω-conotoxin MVIIA at doses that did not impair motor function; the toxin was found to be 100 times more potent than morphine. It seems clear that several other neurological or psychiatric diseases will benefit from the development of drugs that interfere selectively with different Ca^{2+} channel subtypes.

From the molecular genetic point of view, interesting clinical findings start to emerge. For instance, patients with familiar hemiplegic migraine have a defect in their P/Q channel gene (Ophoff et al. 1996). It is interesting that dotarizine and flunarizine, which block these channels (Villarroya et al. 1995), are efficacious in the prophylactic treatment of migraine. P/Q-type Ca^{2+} channel defects have also been found in ataxia and epilepsy (Ophoff et al. 1998). On the other hand, the use of photoaffinity labeling, chimeric α_1 subunits, and site-directed mutagenesis have led to the identification of the amino acids involved in DHP binding to L-type Ca^{2+} channels. It is interesting that the insertion of the drug-binding amino acids enables the transfer of drug sensitivity into Ca^{2+} channels that are insensitive to DHPs (Striessnig et al. 1998). These molecular studies will surely facilitate the design and synthesis of new compounds with affinities for diverse neuronal Ca^{2+} channels, with potential pharmacological and therapeutic benefits in various CNS disorders.

Acknowledgments. Work in our laboratory related to the subject of this review is supported by Fundación Teófilo Hernando, and by grants from DGICYT (PB94-0185 to C.M., and PB94-0150 to AGG) and CAM (0.89/0001/1997).

Reference

Adams ME, Bindokas VP, Hasegawa L, et al (1990) ω-Agatoxins: novel calcium channel antagonists of two subtypes from Funnel web spider (Agelenopsis aperta) venom. J Biol Chem 265:861-867

Ahmad SN, Miljanich GP (1988) The calcium antagonist, ω-conotoxin, and electric organ nerve terminals: binding and inhibition of transmitter release and calcium influx. Brain Res 453:247-256

Ahlijanian MK, Striessnig J, Catterall WA (1991) Phosphorylation of an α_1-like subunit of an ω-conotoxin-sensitive brain Ca^{2+} channel by cAMP-dependent protein kinase and protein kinase C. J Biol Chem 266:20192-20197

Akaike N, Tsuda Y, Oyama Y (1988) Separation of current- and voltage-dependent inactivation of calcium current in frog sensory neuron. Neurosci Lett 84:46-50

Albillos A, Artalejo AR, López MG, et al (1994) Ca^{2+} channel subtypes in cat chromaffin cells. J Physiol 477:197-213

Albillos A, García AG, Gandía L (1993) ω-Agatoxin-IVA-sensitive calcium channels in bovine chromaffin cells. FEBS Lett 336:259-262

Albillos A, García AG, Olivera BM, et al (1996) Re-evaluation of the P/Q Ca^{2+} channel components of Ba^{2+} currents in bovine chromaffin cells superfused with low and high Ba^{2+} solutions. Pflügers Arch 432:1030-1038

Armstrong, D.L. (1989) Ca^{2+} channel regulation by calcineurin, a Ca^{2+}-activated phosphatase in mammalian brain. Trends Neurosci 12:117-122

Artalejo CR, Adams ME, Fox AP (1994) Three types of Ca^{2+} channels trigger secretion with different efficacies in chromaffin cells. Nature 367:72-76

Artalejo CR, Perlman RL, Fox AP (1992) ω-Conotoxin GVIA blocks a Ca^{2+} current in chromaffin cells that is not of the "classic" N Type. Neuron 8:85-95

Bean BP, Nowycky MC, Tsien RW (1984) β-adrenergic modulation of Ca^{2+} channels in frog ventricular heart cells. Nature 307:371-375

Belardinelli L, Isenberg G (1983) Actions of adenosine and isoproterenol on isolated mammalian ventricular myocytes. Circ Res 53:287-297

Bezprozvanny I, Tsien RW (1995) Voltage-dependent blockade of diverse types of voltage-gated Ca channels expressed in Xenopus oocytes by the Ca^{2+} channel antagonist mibefradil (Ro 40-5967). Mol Pharmacol 48:540-549

Bossu J-L, De Waard M, Feltz A (1981) Inactivation characteristics reveal two calcium current in adult bovine chromaffin cells. J Physiol 437:603-620

Bourinet E, Fournier F, Nargeot J, et al (1992) Endogenous *Xenopus*-oocyte Ca^{2+}-channels are regulated by protein kinases A and C. FEBS Lett 299:5-9

Bowman D, Alexander S, Lodge D (1993) Pharmacological characterisation of the calcium channels coupled to the plateau phase of KCl-induced intracellular free Ca^{2+} elevation in chicken and rat synaptosomes. Neuropharmacology 32:1195-1202

Brehm P, Eckert R (1978) Calcium entry leads to inactivation of calcium channels in *Paramecium*. Science 202:1203-1206

Brust PF, Simerson S, McCue AF, et al (1993) Human neuronal voltage-dependent calcium channels: studies on subunit structure and role in channel assembly. Neuropharmacology 32:1089-1102

Cachelin AB, De Peyer JE, Kokubun S, et al (1983) Ca^{2+} channel modulation by 8-bromocyclic AMP in cultured heart cells. Nature 304:462-464 classical T-type currents, although the more slowly inactivating neuronal T currents may represent activation of a mixture of α1 subunits including α1G, as predicted previously (Chen and Hess 1990).

Cytoplasmic β subunits play a significant role in the modulation of the currents generated by individual HVA α1 subunits. In general, co-expression of the β subunits with α1 subunits results in an increase in calcium current amplitude, by increasing expression at the plasma membrane (Chien et al. 1995; Brice et al. 1997). Co-expression of the α2-δ subunit with α1 subunits also causes a potentiation in current, particularly when β is also present (Williams et al. 1992; Brust et al. 1993; Gurnett et al. 1996). However, the role of auxiliary subunits in the modulation of LVA currents is less clear. It has recently been established that the antisense depletion of β subunits in nodose ganglion neurons has no effect on the endogenous T-type currents (Lambert et al. 1997), although native T-type channels in oocytes have been reported to be modulated by β subunits (Lacerda et al. 1994).

In the present paper I review our work on native T-type currents, and our studies on cloned and expressed calcium channels, which have the aim of probing the molecular nature of T-type currents and channels.

1.1 Functional Importance of T-Type Channels in Neuronal Firing

In cells that contain T-type currents, membrane hyperpolarization to deinactivate the T-type currents can produce low threshold spikes, which themselves trigger Na^+ spikes, resulting in a bursting pattern of firing. The pattern of bursting depends on the biophysical properties of the T-type currents (Huguenard and Prince 1992). Furthermore, several groups have suggested that alterations in T-type channels may be of importance in the generation of epileptiform activity (Huguenard 1996; Tsakiridou et al. 1995), and certain pharmaceutical agents of use in anaesthesia and in the treatment of epilepsies, including phenytoin, are known to block T-type currents, within the range of concentrations used clinically (Kobrinsky et al. 1994; Huguenard 1996; Todorovic and Lingle 1998).

2 Results

2.1 Studies on Native T-Type Currents

In published studies (Kobrinsky et al. 1994; Scott et al. 1990), we have investigated the properties of T-type voltage dependent calcium channels in two cell types, dorsal root ganglion neurons (DRGs) and a DRG-neuroblastoma hybrid cell line (ND7-23) (Dunn et al. 1991; Wood et al. 1990). The ND7-23 cell line showed only LVA calcium currents in all undifferentiated cells. In these cells, two types of LVA currents were observed, fast and slowly inactivating (Kobrinsky et al. 1994). The τ_{inact} was voltage dependent for both current types, being about 21 and 55 ms at –40 mV for the rapidly and slowly inactivating currents, respectively (Kobrinsky et al. 1994). The slowly inactivating currents also showed a noninactivating plateau current, which was very long lasting. The slowly inactivating LVA currents predominated in most preparations; for this reason most pharmacology was performed on these currents. Complete block of the slowly inactivating LVA currents was observed by Ni^{2+} (100 μM), and almost complete (84%) but rapidly reversible block was achieved by ω-conotoxin GVIA (1 μM) (Kobrinsky et al., 1994). By contrast with studies on T-type currents from some central neurons (Akaike et al, 1989), we observed no effect of the 1,4-dihydropyridine antagonists (-)-202-791 or isradipine (Kobrinsky et al. 1994), although these drugs were not examined on the rapidly inactivating T-type current, which is the subtype of LVA current selectively inhibited by DHPs in some central neurons (Akaike et al. 1989; Tarasenko et al. 1997). Phenytoin at a concentration of 10 μM inhibited about 30% of the T-type current in ND7-23 cells, and at this concentration had no effect on HVA current.

It would appear from a comparison of our own and other pharmacological studies that T-type currents represent a disparate group of currents rather than a single entity (Huguenard 1996). For example, for the T-type currents in the 3T3 epithelial cell line, the voltage-independent minimum of the τ_{inact} was less than 10 ms, whereas for NG108-15 cells, this was about 30 ms (Chen and Hess 1990). Furthermore, for the slowly inactivating T-type currents in rat thalamic neurons, the τ_{inact} was between 50 and 70 ms (Huguenard and Prince, 1992; Tarasenko et al. 1997), whereas the rapidly inactivating currents showed τ_{inact} values of about 30 ms. In several cell types, for example NG108-15 cells there may be a mixed population of rapidly and slowly inactivating T-type currents (Chen and Hess 1990). The rapidly inactivating α1G currents showed a voltage-dependent minimum τ_{inact} of about 10 ms (Perez-Reyes et al. 1998), equating them in this respect with the rapidly inactivating epithelial cell T current (Chen and Hess 1990).

2.2 Comparison of the Properties of T-Type Current and Cloned α1E Currents

It has been suggested that the α1E VDCC clone corresponds to an LVA current and exhibits many of the properties shown by native T-type currents (Soong et al., 1993; Bourinet et al. 1996). We have examined the properties of rat α1E (rbEII) transiently transfected in COS-7 cells and co-expressed with the brain α2-δ splice variant and β1b (Stephens et al. 1997). The current inactivation kinetics were even slower even than the slowly inactivating T-type current. For example, the τ_{inact} at 0mV was about 150 ms. In the absence of co-expressed accessory subunits, the inactivation rate was more rapid, but not as rapidly inactivating as for the range of

Chen C, Hess P (1990) Mechanism of gating of T-type calcium channels. J Gen Physiol 96:603-630

Chien AJ, Zhao XL, Shirokov RE, et al (1995) Roles of a membrane-localized β subunit in the formation and targeting of functional L-type Ca^{2+} channels. J Biol Chem 270:30036-30044

Dubel SJ, Starr TVB, Hell J, et al (1992) Molecular cloning of the α-1 subunit of an ω-conotoxin-sensitive calcium channel. Proc Natl Acad Sci USA 89:5058-5062

Dunn PM, Coote PR, Wood JN, et al (1991) Bradykinin evoked depolarization of a novel neuroblastoma x DRG neurone hybrid cell line (ND7-23). Brain Res. 549:80-86

Ellis SB, Williams ME, Ways NR, et al (1988) Sequence and expression of mRNAs encoding the α$_1$ and α$_2$ subunits of a DHP-sensitive calcium channel. Science 241:1661-1664

Fedulova SA, Kostyuk PG, Veselovsky NS (1985) Two types of calcium channels in the somatic membrane of new-born rat dorsal root ganglion neurones. J Physiol (Lond) 359:431-446

Gurnett CA, De Waard M, Campbell KP (1996) Dual function of the voltage-dependent Ca^{2+} channel α$_2$δ subunit in current stimulation and subunit interaction. Neuron 16: 431-440

Herrington J, Lingle CJ (1992) Kinetic and pharmacological properties of low voltage activated Ca^{2+} current in rat clonal GH$_3$ pituitary cells. J Neurophysiol 68:213-231

Huguenard JR (1996) Low-threshold calcium currents in central nervous system neurons. Annu Rev Physiol 58:329-348

Huguenard JR, Prince DA (1992) A novel T-type current underlies prolonged Ca^{2+}-dependent burst firing in GABAergic neurons of rat thalamic reticular nucleus. J Neurosciences 12:3804-3817

Kobrinsky EM, Pearson HA, Dolphin AC (1994) Low- and high-voltage-activated calcium channel currents and their modulation in the dorsal root ganglion cell line ND7-23. Neuroscience 58:539-552

Lacerda, A.E., Perez-Reyes, E., Wei, X., et al (1994). T-type and N-type calcium channels of *Xenopus* oocytes: evidence for specific interactions with β subunits. Biophys J 66:1833-1843

Lambert RC, Maulet Y, Mouton J, et al (1997) T-type Ca^{2+} current properties are not modified by Ca^{2+} channel β subunit depletion in nodosus ganglion neurons. J Neurosci 17:6621-6628

Meir A, Dolphin AC (1998) Known calcium channel α1 subunits can form low threshold, small conductance channels, with similarities to native T type channels. Neuron 20:341-351

Nilius B, Hess P, Lansman JB, Tsien RW (1985) A novel type of cardiac calcium channel in ventricular cells. Nature 316:443-446

Perez-Reyes E, Castellano A, Kim HS, et al (1992) Cloning and expression of a cardiac/brain β subunit of the L-type calcium channel. J Biol Chem 267:1792-1797

Perez-Reyes E, Cribbs LL, Daud A, Lacerda AE, Barclay J, Williamson MP, Fox M, Rees M, Lee J (1998). Molecular characterisation of a neuronal low-voltage-activated T type calcium channel. Nature 391:896-900

Perez-Reyes E, Schneider T (1994) Calcium channels: structure, function, and classification. Drug Dev Res 33:295-318

Pragnell M, Sakamoto J, Jay SD, et al (1991) Cloning and tissue-specific expression of the brain calcium channel β-subunit. FEBS Lett 291:253-258

Schneider T, Wei X, Olcese R, et al. (1994). Molecular analysis and functional expression of the human type E neuronal Ca^{2+} channel α1 subunit. Receptors and Channels 2:255-270

Scott RH, Wootton JF, Dolphin AC (1990) Modulation of neuronal T-type calcium channel currents by photoactivation of intracellular guanosine 5'-O(3-thio) triphosphate. Neuroscience 38:285-294

Soong TW, Stea A, Hodson CD, et al (1993). Structure and functional expression of a member of the low voltage-activated calcium channel family. Science 260:1133-1136

Stephens GJ, Page KM, Burley JR, et al (1997) Functional expression of rat brain cloned α1E calcium channels in COS-7 cells. Pflügers Arch 433:523-532

Tanabe T, Takeshima H, Mikami A, et al(1987) Primary structure of the receptor for calcium channel blockers from skeletal muscle. Nature 328:313-318

Tarasenko AN, Kostyuk PG, Eremin AV, et al (1997) Two types of low-voltage-activated Ca^{2+} channels in neurones of rat laterodorsal thalamic nucleus. J Physiol (Lond) 499:77-86

Todorovic SM, Lingle CJ (1998) Pharmacological properties of T-type Ca^{2+} current in adult rat sensory neurons: effects of anticonvulsant and anesthetic agents. J Neurophysiol 79:240-252

Tsakiridou E, Bertollini L, De Curtis M, et al (1995) Selective increase in T-type calcium conductance of reticular thalamic neurons in a rat model of absence epilepsy. J. Neurosci. 15:3110-3117

Williams ME, Feldman DH, McCue AF, et al (1992) Structure and functional expression of α$_1$, α$_2$, and β subunits of a novel human neuronal calcium channel subtype. Neuron 8:71-84

Williams ME, Marubio LM, Deal CR, et al (1994) Structure and functional characterization of neuronal α$_{1E}$ calcium channel subtypes. J Biol Chem 269:22347-22357

Wood JN, Bevan SJ, Coote PR, et al (1990) Novel cell lines display properties of nociceptive sensory neurons. Proc R Soc Lond [B] 241:187-194

Wyatt CN, Pagee KM, Berrow NS, et al. (1998) The effect of overexpression of auxiliary calcium channel subunits on native Ca^{2+} channel currents in undifferentiated NG-108-16 cells. J Physiol 510:347-360

Exocytosis Calcium Channels: Autocrine/Paracrine Modulation

A.G. GARCÍA[1,2], J.M. HERNÁNDEZ-GUIJO[1], I. MAYORGAS[1], and L.GANDÍA[1]

[1]Instituto de Farmacología Teófilo Hernando, Departamento de Farmacología, Facultad de Medicina, Universidad Autónoma de Madrid, Arzobispo Morcillo 4, 28029 Madrid, Spain
[2]Servicio de Farmacología Clínica e Instituto de Gerontología, Hospital Universitario de la Princesa, Diego de León, 62, 28006 Madrid, Spain

Key words. Calcium channels, Chromaffin cells, Purinergic receptors, Opiate receptors, Catecholamines, Facilitation

1 Introduction

This review focuses on the diversity of the high-voltage-activated (HVA) Ca^{2+}channels expressed by adrenal medullary chromaffin cells of various animal species, and on their regulation by endogenous materials coreleased with the catecholamines noradrenaline and adrenaline. From our point of view, this autocrine/paracrine modulation constitutes an adequate framework to explain the so-called facilitation of Ca^{2+} channel current, a phenomenom first described in the laboratory of Erwin Neher in bovine chromaffin cells (Fenwick et al. 1982). Facilitation consists of the augmentation of the whole-cell inward current flowing through Ca^{2+} channels, when a test depolarizing pulse is preceded by a strong depolarizing prepulse. This Ca^{2+} current facilitation may be the basis for the regulation of adrenal medullary catecholamine release during stress, as well as for the modulation of neurotransmitter release mediated by presynaptic autoreceptors.

In a restricted space such as the synapse, the modulation of the rate and the amounts of action potential-triggered neurotransmitter release seems to be controlled through a negative feedback mechanism mediated by presynaptic autoreceptors (Kirpekar and Puig 1971). These receptors are activated by their own neurotransmitter that, at least in sympathetic neurones, acts on presynaptic α_2 receptors to inhibit noradrenaline release. Such inhibition seems to be related to blockade of N-type Ca^{2+} channels, as suggested from experiments measuring Ca^{2+} channel currents and K^+-evoked noradrenaline release from sympathetic neurones. Direct access to sympathetic nerve terminals is not possible, and thus those electrophysiological and neurosecretory experiments provided only indirect evidence to reach such assumptions (Lipscombe et al. 1989). Although adrenal medullary chromaffin cells are close relatives of sympathetic neurones both embryologically and functionally, they however do not have an α_2-mediated feedback mechanism for the regulation of noradrenaline and adrenaline release (Powis and Baker 1986; Orts et al. 1987). In spite of this, chromaffin cells have become an excellent model to study the modulation of neurotransmitter release by autoreceptors because (a) they express several HVA Ca^{2+} channel subtypes with a pharmacology similar to that of neuronal Ca^{2+} channels (Olivera et al. 1994; García et al. 1996); (b) the current flowing through these channels suffers a drastic modulation by ATP and opiate receptors; (c) chromaffin cells exhibit a potent, well-characterized, and easily measurable catecholamine exocytotic response. Thus, like sympathetic neurones these cells may have a mechanism to control the release of catecholamines, with a target on HVA Ca^{2+} channels; but unlike sympathetic neurones, P_{2y} purinergic receptors and μ and δ opiate receptors (but not α_2 adrenergic receptors) coupled to G proteins and Ca^{2+} channels seem to control the access of external Ca^{2+} to the secretory machinery. This review focuses on the experimental basis supporting the existence of such a regulatory mechanism in adrenal medullary chromaffin cells.

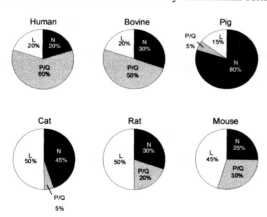

Fig. 1. Different proportions of high-voltage-activated (HVA) Ca^{2+} channels are expressed in adrenal medullary chromaffin cells of the six animal species so far studied using patch-clamp techniques. (See García et al. 1996 for further details.) Data for pig chromaffin cells are taken from Kitamura et al. (1997).

2 Diversity of Ca^{2+} Channels Expressed by Chromaffin Cells: Drastic Species Differences

Chromaffin cells express a variety of HVA Ca^{2+} channels, but neither R-type channels nor low-voltage-activated (LVA; T-type) Ca^{2+} channels are found. The relative density of each channel subtype (L, N, P/Q) has been determined through quantitation of the fraction of whole-cell current through Ca^{2+} channels that is inhibited by various 1,4-dihydropyridines or ω-toxins. Table 1 summarizes the

In more recent studies, Doupnik and Pun (1994), Albillos et al. (1996b), and Currie and Fox (1996) observed that the rate of activation and the amplitude of Ba^{2+} currents in bovine chromaffin cells depend critically on the experimental superfusion conditions of the patch-clamped cell. Cell activity under flow-stop conditions (unperfused cell) favor the local rise of secreted products outside the plasmalemma, the subsequent activation of membrane autoreceptors, and the rapid inhibition of spatially localized Ca^{2+} channels. This "tonic" inhibition, induced by low molecular weight compounds of the vesicle content (i.e., ATP and opiates), coreleased with the catecholamines during application of depolarizing pulses under flow-stop conditions, can be greatly reversed ("facilitated") by strong depolarizing pre-pulses. The tonic inhibition of the current is also reversed on resuming the rapid flow over the surface of the cell to quickly wash out   the released materials (Fig. 4C). If 10 mM Ca^{2+} (instead of 10 mM Ba^{2+}) was used as a charge carrier, flow-stop caused a smaller modulation of the current (Fig. 4D). If in addition to using 10 mM Ca^{2+} as a charge carrier the cell was dialyzed with BAPTA (a strong and rapid Ca^{2+} chelator), the Ca^{2+} channel current was similar under flow or flow-stop conditions (Fig. 4E). This was likely caused by the rapid sequestration of Ca^{2+} near exocytotic sites and the suppression of secretion.

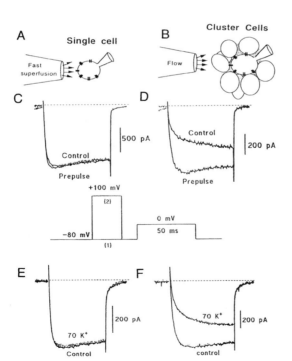

Fig. 5A-F. Modulation of Ca^{2+} channel currents in a single cell (**A**, **C**, **E**) or in a cell immersed in a cell cluster (**B**, **D**, **F**). Cells **C** and **D** were superfused with an extracellular solution containing 10 mM Ba^{2+}; cells **E** and **F** were superfused with 10 mM Ca^{2+}. Before the test pulse to 0 mV (with or without prepulse to +100 mV), a 70 mM K^+ solution was applied during 1 s (70 K^+). (See text for further details).

8 A cell in Isolation Versus a Cell Inmersed in a Cluster Have Different Ca²⁺ Channel Current Modulation

When plated at higher densities, bovine chromaffin cells tend to cluster to form a kind of "islet." So, Ca^{2+} channel currents can be recorded from one voltage-clamped cell while the surrounding cells are stimulated to release materials (i.e. ATP, opiates) that bathing the test cell will modulate its Ca^{2+} channel currents. This experimental situation reproduces closely the physiological arrangement of chromaffin cells in the intact adrenal medullary tissue.

Figure 5D shows a pair of current traces obtained from a voltage-clamped cell that was immersed in a cell cluster (panel B). All cells in the cluster were continuously superfused by an extracellular solution containning 10 mM Ba^{2+}, a situation that will cause a strong secretory response from these unclamped cells, as the experiment of Fig. 3A suggests. As a consequence, the first Ba^{2+} current trace obtained without a prepulse exhibited a very slow activation and a marked reduction of its amplitude (control trace in panel D). The prepulse accelerated the current activation and increased its amplitude In a similar experiment performed in a single isolated cell, control current showed rapid activation and the prepulse caused no current facilitation (panel D).

The experiments shown in Fig.5E were performed in cells superfused with 10 mM Ca^{2+} (instead of Ba^{2+}). As suggested by the experiment in Fig. 3, Ca^{2+} ions require K^+ depolarization to activate exocytosis in chromaffin cells. So, in panel F, secretion in a cell cluster was stimulated with a brief application of a solution enriched in K^+ (70 mM K^+; 1 s); under these conditions, the I_{Ca} recorded from the patch-clamped cell activated with a slow kinetics and showed a clear facilitation by strong depolarizing prepulses, suggesting that the voltage-clamped cell is under the modulatory influence of materials released from neighboring cells by the K^+ pulse. Again, in an isolated single cell, modulation of I_{Ca} (after the K^+ pulse) could not be seen (panel E).

9   Modulation of L-Type Versus N/P/Q Type Ca²⁺ Channels

Because chromaffin cells express at least three types of HVA Ca^{2+} channels (L, N, P/Q), the question is whether all of these are equally modulated by endogenously released ATP and opiates. It is interesting that L- and non-L-type Ca^{2+} channels (N, P/Q) of bovine chromaffin cells seem to be modulated by opioids through different mechanisms (Albillos et al. 1996a). L channels are modulated mostly through a voltage-independent mechanism (no facilitation by prepulses was observed associated to L channels); non-L channels appear to be inhibited through a voltage-dependent pathway. Thus, nifedipine (an L-type channel blocker; Table 1) blocks the voltage-independent component with high selectivity (60% block of the voltage-independent component, 5% block of the voltage-dependent). On the other hand, ω-conotoxin MVIIC (a blocker of P/Q channels), blocks completely the voltage-dependent component. This voltage-dependent component was distributed 58% for N channels and 42% for P/Q channels. These data are similar to those recently obtained by Currie and Fox (1997), also in bovine chromaffin cells, but using ATP to modulate the Ca^{2+} channel currents: N-type channels contribute 69% and P/Q channels 47% to the voltage-dependent modulation.

Facilitation of Ca^{2+} channel currents by depolarizing prepulses was blocked by ω-conotoxin GVIA in cat chromaffin cells (Albillos et al. 1994) and by ω-conotoxin MVIIC in bovine chromaffin cells (Albillos et al. 1996a). Dihydropyridines did not affect such facilitation, suggesting that the N/P/Q- but not L subtypes of Ca^{2+} channels are involved in the voltage-dependent modulation/facilitation of Ca^{2+} channel currents. Contrary to our results, in a series of six papers from Aaron Fox's laboratory (Artalejo et al. 1990; 1991a,b; 1992a,b, 1994), it is shown that prepulse facilitation of Ca^{2+} channel currents is blocked entirely by nisoldipine (an L-type channel blocker). In addition, cell dialysis with GTP-γS or GDP-βS had no effects on L-channel-associated facilitation, either by prepulses or cAMP (Artalejo et al. 1992b). In a recent debate, Cristina R. Artalejo, one of the co-authors of the six papers from Fox's laboratory, argued that "the expression of the facilitation L channel is strongly dependent on the age of the animals from which chromaffin cells are prepared" (Artalejo 1997). Artalejo suggests that the discrepancy between our results (facilitation is mainly associated to N/P/Q-type Ca^{2+} channels) and her results with Fox (facilitation is associated to L-type channels) results because we used adult bovine animals while they used 10- to 12-week-old calves.

What seems most surprising is that, after six papers on the existence of the L-type facilitation current in bovine chromaffin cells, a modulatory pathway associated to N/P/Q-type Ca^{2+} channels is reported by Aaron Fox's laboratory in two recent papers (Currie and Fox 1996, 1997). In frank coincidence with our earlier data with ATP (Gandía et al. 1993), 3 and 4 years later Fox concludes (as we did) that ATP modulates the non-L-type Ca^{2+} channels through a G-protein-coupled pathway in bovine chromaffin cells. It is puzzling that there is no mention of the age of the animals used as donors of adrenal glands in these two papers. It is also curious that all experiments reported in these two papers were done in the presence of nisoldipine, to avoid (according to the authors) obscuring their results by the facilitation associated to L-type channels. Would it not be adequate to include experiments using ω-conotoxin MVIIC (to block N/P/Q-channels) and see whether facilitation of L channels occurred under these conditions? We could not show any prepulse facilitation associated to L channels in our chromaffin cells when N/P/Q Ca^{2+} channels were blocked by ω-toxins (Albillos et al. 1996b). If done in Fox's laboratory, this simple experiment could easily solve the issue of whether "baby" calves have both types of facilitation, one associated to L-type channels and the other to N/P/Q-type channels.

We believe that the reasons for the earlier contrasting results may have different origins and rely on the complexity of the chromaffin cell system, the autocrine/paracrine nature of Ca^{2+} channel modulation (Gandía et al. 1993; Doupnik and Pun 1994; Albillos et al. 1996b), and the possible coupling between channel subunits and the second-messenger pathways affecting the up- or downregulation of Ca^{2+} channel subtypes (Hille 1994). We do not exclude the possibility that L-type and non-L-type channel facilitation may coexist in the same cell and that, depending on various conditions (for example, superfusion versus flow-stop of a cell in isolation versus a cell inmersed in a cell cluster, a more physiological situation), one or the other may be overlooked or overstressed. However, this issue is complex and should be approached with care (Carbone and García 1997). Here, we would like to point out once again that the membrane-delimited N/P/Q-type channel modulated by voltage and G proteins present in neurones (Hille 1994) is a well-established phenomenon in chromaffin cells, highly reproducible, and now recognized by six independent laboratories at the macroscopic (Gandía et al. 1993; Doupnik and Pun 1994; Albillos et al. 1994, 1996b; Currie and Fox 1996; Kitamura et al. 1997) and single channel level (Carabelli et al 1996). The same cannot be concluded for the L-type facilitation current reported by Artalejo and Fox, which among chromaffin cells seems to be confined to young calves. A recent article on Ca^{2+} currents in rat chromaffin cells (Hollins and Ikeda 1996) described a voltage-dependent facilitation of L-type channels that deviates significantly from that observed in Fox's laboratory for the bovine cells. In the rat, facilitation associated to L-type channels contributes at most to 6% of the total current; it is insensitive to D_1 dopamine agonists and to PKA activation, and is short lasting rather than persisting for seconds. Thus, there seem to be enough reasons to stop arguing and try to find a rationale for these increasing discrepancies.

10 Functional Significance of Ca^{2+} Channel Facilitation for the Regulation of Catecholamine Release

The profound slowing down of current activation and the inhibition of peak Ca^{2+} channel currents by ATP and opiates must cause profound alterations of Ca^{2+} entry, local $[Ca^{2+}]_i$ transients at exocytotic subplasmalemmal sites, and of the rates of noradrenaline and adrenaline release. This may constitute the basis for the fine tuning of the amount of adrenal medullary catecholamines, delivered to the circulation during various stressful conflicts. This fine regulation is strictly needed to prevent massive uncontrolled release of catecholamines, which could lead to hypertensive crisis, arrhythmias, and myocardial damage. Catecholamines are needed quickly and at adequate concentrations, in target organs needing their adaptation to mental or physical stress or to emergences requiring a fast "fight-or-flight" response of the entire body. But catecholamines stored in the two adrenal glands can cause death of the animal if released suddenly into the circulation: a precise and efficient control of their rate of secretion is needed. With the data here reported, we envision the regulation of catecholamine release to the circulation as presented in the scheme of Fig. 6.

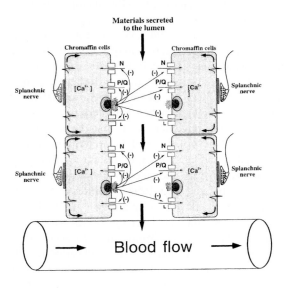

Fig. 6. Scheme showing our present view on the modulation of adrenal medullary Ca^{2+} channels and catecholamine release by an autocrine/para-crine feedback mechanism acti-vated by ATP and opiates coreleased with adrenaline and noradrenaline.

In the adrenal medulla of mammals, chromaffin cells adopt a columnar disposition around a small capillary vessel; the cells in the intact gland are believed to be polarized (Carmichael 1986). Their secretory surface is exposed to the concentrations of secretory materials released from the cells (i.e., catecholamines, ATP, opiates). In resting conditions, the splanchnic nerves fire action potentials at low frequency (less than 1 Hz); as a result, acetylcholine is released into the cholino-chromaffin synapse. This causes cell depolarization, Ca^{2+} channel opening, Ca^{2+} entry, catecholamine release, and a concomitant elevation of opiates and ATP in the immediate vicinity of the cell secretory surface. The activation of P_{2y}, μ, and δ receptors from the same cell (autocrine modulation) or from neighboring cells (paracrine modulation) causes inhibition of Ca^{2+} channels, Ca^{2+} entry, and catecholamine secretion (tonic inhibition). During a sudden stress, the splanchnic nerve markedly increases the firing of action potentials (up to 10-20 Hz); the repetitive acetylcholine-evoked action potentials in the postsynaptic cell relieve the tonic inhibition of Ca^{2+} channels, as suggested by recent elegant experiments performed in cholinergic neurones (Williams et al. 1997). Relieve of the tonic inhibition (facilitation) will secure the enhancement of Ca^{2+} entry required for the massive release of catecholamines during the "fight-or-flight" response, or during severe disease conditions, i.e., asthma crisis, acute myocardial infarction, anaphylactic shock, or heart failure, in which a strong sympathoadrenal stimulation exists. The fact that this modulatory mechanism is present also in human chromaffin cells (Gandía, Mayorgas and García, 1998) reinforces its physiological and pathophysiological relevance. Thus, fine tuning of the rate of catecholamine release in each moment will be achieved through the interplay between the degree of tonic activation of P_{2y} and μ and δ autoreceptors by endogenous agonists coreleased with the catecholamines and the rate of action potential firing in chromaffin cells, controlled by the rate of firing of splanchnic nerve terminals, the release of acetylcholine, and the activation of nicotinic receptors on the surface of chromaffin cells.

11 Conclusions and Perspectives

The use of flow conditions or static incubation (flow-stop) has surely been the cause of several misinterpretations of the mechanisms underlying Ca^{2+} channel facilitation (García and Carbone 1996; Carbone and García 1997) since its discovery in Erwin Neher's laboratory (Fenwick et al. 1982). This is particularly true for cells with a powerful secretory response such as chromaffin cells, where flow or flow-stop conditions or the use of Ca^{2+} or Ba^{2+} as charge carriers might deeply modify the rate and quantities of secretory materials that will bathe the cell surface.

There is a lack of direct evidence showing that the modulation of Ca^{2+} channels might control the rate of catecholamine release. As far as we know, there is only a recent report showing that exogenous ATP inhibits Ca^{2+} current and exocytosis (measured with capacitance techniques) in the same chromaffin cell (Lim et al. 1997). However, to understand the significance of this regulatory process at a more physiological level, experiments similar to those reported here (acetylcholine stimulation of cell clusters, flow-stop), simultaneously measuring Ca^{2+} channel currents and exocytosis, should be carried out. In addition, experiments in adrenal medullary slices (Moser and Neher 1997) where the relationship between cells and their innervation by splanchnic nerve terminals is preserved are also critical to understand the physiological modulation of Ca^{2+} channel current and secretion.

Acknowledgements. This study was supported by grants from DGICYT (PB94-0150), CAM (08.9/0001/1997), and Fundación Teófilo Hernando. We thank M.C. Molinos for the typing this manuscript.

References

Albillos A, Artalejo AR, López MG, et al (1994) Ca^{2+} channel subtypes in cat chromaffin cells. J Physiol 477:197-213

Albillos A, Carbone E, Gandía L, et al (1996a) Opioid inhibition of Ca²⁺ channel subtypes in bovine chromaffin cells: selectivity of action and voltage-dependence. Eur J Neurosci 8:1561-1570

Albillos A, Gandía L, Michelena P, et al (1996b) The mechanism of calcium channel facilitation in bovine chromaffin cells. J Physiol 494:687-695

Albillos A, García AG, Olivera B, et al (1996c). Re-evaluation of the P/Q Ca²⁺ channel component of Ba²⁺ currents in bovine chromaffin cells superfused with low and high Ba²⁺ solutions. Pflügers Arch 432:1030-1038

Amico C, Marchetti C, Nobile M, et al (1995) Pharmacological types of Ca²⁺ channels and their modulation by baclofen in cerebellar granules. J Neurosci 15:2839-2848

Artalejo CR (1997) More on calcium currents. Trends Neurosci 20:448-449

Artalejo CR, Ariano MA, Perlman RL, et al (1990) Activation of facilitation calcium channels in chromaffin cells by D₁ dopamine receptors through a cAMP/protein kinase A-dependent mechanisms. Nature 384:239-247

Artalejo CR, Dahmer MK, Perlman RL, et al (1991a) Two types of Ca²⁺ currents are found in bovine chromaffin cells: facilitation is due to the recruitment of one type. J Physiol 432:681-707

Artalejo CR, Dahmer MK, Perlman RL, et al (1991b) Three types of bovine chromaffin cell Ca²⁺ channels: facilitation increases the open probability of a 27 pS channel. J. Physiol 444:213-240

Artalejo CR, Mogul DJ, Perlman RL, et al (1994) Three types of Ca²⁺ channels trigger secretion with different efficacies in chromaffin cells. Nature 367:72-76

Artalejo CR, Perlman RL, Fox AP (1992a) ω-Conotoxin GVIA blocks a Ca²⁺ current in chromaffin cells that is not of the "classic" N type. Neuron 8:85-95

Artalejo CR, Perlman RL, Fox AP (1992b) Voltage-dependent phosphorylation may recruit Ca²⁺ current facilitation in chromaffin cells. Nature 358:63-66

Bley KR, Tsien RW (1990) Inhibition of Ca²⁺ and K⁺ channels in sympathetic neurons by neuropeptides and other ganglionic transmitters. Neuron 4:379-391

Boland LM, Bean B P (1993) Modulation of N-type Ca²⁺ channels in bullfrog sympathetic neurons by luteinizing hormone-releasing hormone: kinetics and voltage dependence. J Neurosci 13:516-533

Borges R, Sala F, García AG (1986) Continuous monitoring of catecholamine release from perfused cat adrenals. J Neurosci Methods 16:389-400

Callewaert G, Johnson RG, Morad M (1991) Regulation of the secretory response in bovine chromaffin cells. Am J Physiol 260:C851-C860

Carabelli V, Lovallo M, Magnelli V, et al (1996) Voltage-dependent modulation of single N-type Ca²⁺ channel kinetics by receptor agonists in IMR32 cells. Biophys J 70:2144-2154

Carbone E, García AG (1997) More on calcium currents. Trends Neurosci 20:448-450

Carmichael SW (1986) Morphology and innervation of the adrenal medulla. In: Rosenheck K, Lelkes P (eds) Stimulus-secretion coupling, vol 1. CRC Press, Boca Ratón, Rosenheck, K., Lelkes, P. (eds) pp 1-29

Cox DH, Dunlap K (1992) Pharmacological discrimination of N-type from L-type Ca²⁺ current and its selective modulation by transmitters. J Neurosci 12:906-914

Currie KPM and Fox AP (1996) ATP serves as a negative feed-back inhibitor of voltage-dependent Ca²⁺ channel currents in cultured bovine adrenal chromaffin cells. Neuron 16:1027-1036

Currie KPM, Fox AP (1997) Comparison of N- and P/Q-type voltage-dependent calcium current inhibition. J Neurosci 17:4570-4579

Deisz RA, Lux HD (1985) γ-Aminobutyric acid-induced depression of Ca^{2+} currents of chick sensory neurons. Neurosci Lett 56:205-210

Dolphin AC (1996) Facilitation of Ca^{2+} current in excitable cells. Trends Neurosci 19:35-43

Dolphin AC (1998) Mechanisms of modulation of voltage-dependent calcium channels by G proteins. J Physiol 506:3-11

Dolphin AC, Scott RH (1987) Ca^{2+} channel currents and their inhibition by (-)baclofen in rat sensory neurones, modulation by guanine nucleotides J. Physiol 386:670-672

Doupnik CA, Pun RYK (1994) G-protein activation mediates prepulse facilitation of Ca^{2+} channel currents in bovine chromaffin cells. J Membr Biol 140:47-56

Dunlap K, Fischbach GD (1978) Neurotransmitters decrease the Ca^{2+} component of sensory neurone action potentials. Nature 276:837-838

Dunlap K, Fischbach GD (1981) Neurotransmitters decrease the Ca^{2+} conductance activated by depolarization of embryonic chick sensory neurones. J Physiol 317:519-535

Fenwick EM, Marty A, Neher E (1982) Sodium and calcium channels in bovine chromaffin cells. J Physiol 331:599-635

Galvan M, Adams PR (1982) Control of Ca^{2+} current in rat sympathetic neurons by norepinephrine. Brain Res 244:135-144

Gandía L, García AG, Morad M (1993) ATP modulation of calcium channels in bovine chromaffin cells. J Physiol 470:55-72

Gandía L, Lara B, Imperial J, et al (1997) Analogies and differences between ω-conotoxins MVIIC and MVIID: binding sites and functions in bovine chromaffin cells. Pflügers Arch 435:55-64

Gandia L, Mayorgar I, Michelena P, et al. (1998) Human adrenal chromaffin cell calcium channels: drastic current facilitation in cell clusters, but not in isolated cells. Pflügers Archiv 436:696-704

García AG, Albillos A, Gandía L, et al (1996) ω-toxins, calcium channels and neurosecretion. In: Gutman Y, Lazarovici P (eds) Cellular and molecular mechanisms of toxin action: toxins and signal transduction. Harwood, Switzerland, pp 155-209

García A, Carbone E (1996) Facilitation of calcium channels. Trends Neurosci 19:383-384

Hescheler J, Rosenthal W, Trautwein W, et al (1987) The GTP-binding protein, G_o, regulates neuronal Ca^{2+} channels. Nature 325:445-447

Hille B (1994) Modulation of ion-channel function by G-protein-coupled receptors. Trends Neurosci 17:531-536

Hirata H, Albillos A, Fernández F, et al (1997) ω-Conotoxins block neurotransmission in the rat vas deferens by binding to different presynaptic sites on the N-type Ca^{2+} channel. Eur J Pharmacol 321:217-223

Hollins B, Ikeda SR (1996) Inward currents underlying action potentials in rat adrenal chromaffin cells. J Neurophysiol 76:1195-1211

Holz GG, Rane SG, Dunlap K (1986) GTP-binding protein mediate transmitter inhibition of voltage-dependent Ca^{2+} channels. Nature 319:670-672

Ikeda SR (1991) Double-pulse calcium current facilitation in adult rat sympathetic neurones. J Physiol 439:181-214

Kirpekar SM, Puig, M (1971) Effects of flow-stop on noradrenaline release from normal spleens and spleens treated with cocaine, phentolamine or phenoxybenzamine. Br J Pharmacol 43:359-369

Kitamura N, Ohta T, Ito S, et al (1997) Calcium channel subtypes in porcine adrenal chromaffin cells. Pflügers Arch 434:179-187

Lewis D, Weight FF, Luini, A (1986) A guanine nucleotide-binding protein mediates the inhibition of voltage dependent Ca^{2+} current by somatostatin in a pituitary cell line. Proc. Natl. Acad. Sci. USA 83:9035-9039

Lim W, Kim SJ, Yan HD, et al (1997) Ca^{2+}-channel-dependent and -independent inhibition of exocytosis by extracellular ATP in voltage-clamped rat adrenal chromaffin cells. Pflügers Arch Eur. J Physiol 435: 34-42

Lipscombe D, Kongsamut S, Tsien RW (1989) α-Adrenergic inhibition of sympathetic neurotransmitter release mediated by modulation of N-type Ca^{2+}-channel gating. Nature 340:639-642

Luebke JI, Dunlap K (1994) Sensory neuron N-type Ca^{2+} currents are inhibited by both voltage-dependent and -independent mechanisms. Pflügers Arch 365:258-262

Marchetti C, Carbone E, Lux HD (1986) Effects of dopamine and noradrenaline on Ca^{2+} channels of cultured sensory and sympathetic neurons of chick. Pflügers Arch 406:104-111

Mintz IM, Bean BP (1993) $GABA_B$ receptor inhibition of P-type Ca^{2+} channels in central neurons. Neuron 10:889-898

Moser T, Neher E (1997) Rapid exocytosis in single chromaffin cells recorded from mouse adrenal slices. J Neurosci 17:2314-2323

Olivera BM, Miljanich G, Ramachandran J, et al (1994) Calcium channel diversity and neurotransmitter release: the ω-conotoxins and ω-agatoxins. Annu Rev Biochem 63:823-867

Orts A, Orellana C, Cantó T, et al (1987) Inhibition of adrenomedullary catecholamine release by propranolol isomers and clonidine involving mechanisms unrelated to adrenoceptors. Br J Pharmacol. 92:795-801

Plummer MR, Rittenhouse A, Kanevsky M, et al (1991) Neurotransmitter modulation of calcium channels in rat sympathetic neurones. J Neurosci 11:2334-2348

Pollo A, Lovallo M, Biancardi E, et al (1993) Sensitivity to dihydropyridines, ω-conotoxin and noradrenaline reveals multiple high-voltage activated Ca^{2+} channels in rat insulinoma and human pancreatic β-cells. Pflügers Arch 423:462-471

Pollo A, Lovallo M, Sher E, et al (1992) Voltage-dependent noradrenergic modulation of ω-conotoxin-sensitive Ca^{2+} channels in human neuroblastoma IMR32 cells. Pflügers Arch Eur J Physiol 422:75-83

Powis DA, Baker PF (1986). α_2 adrenoceptors do not regulate catecholamine secretion by bovine adrenal medullary cells: a study with nicotine. Mol Pharmacol 29:134-141

Swartz KJ (1993) Modulation of Ca^{2+} channels by protein kinase C in rat central and peripheral neurons: disruption of G-protein-mediated inhibition. Neuron 11:305-320

Wanke E, Ferroni A, Malgaroli A, et al (1987) Activation of a muscarinic receptor selectively inhibits a rapidly inactivated Ca^{2+} current in rat sympathetic neurons. Proc Natl Acad Sci USA 84:4313-4317

Williams S, Serafin M, Mühlethaler M, et al (1997) Facilitation of N-type calcium current is dependent on the frequency of action potential-like depolarizations in dissociated cholinergic basal forebrain neurons of the guinea pig. J Neurosci 17:1625-1632

Winkler H, Sietzen M, Schober M (1987) The life cycle of catecholamine-storing vesicles. Ann NY Acad Sci 493:3-19

Synaptic Modulation Mediated by G-Protein-Coupled Presynaptic Receptors

T. TAKAHASHI

Department of Neurophysiology, University of Tokyo Faculty of Medicine
Tokyo 113-0033, Japan

Key words. Presynaptic inhibition, G-protein, Metabotropic glutamate receptor, GABA$_B$ receptor, Synaptic modulation, Long-term depression, EPSC, Slice, Patch-clamp,Whole-cell recording, Calcium channels, GIRK, Potassium channel, Exocytotic machinery, Auditory synapse

1 Synaptic Modulation

Synaptic transmission can be modulated by a variety of factors and for various lengths of time ranging from milliseconds to years. The efficacy of synaptic transmission can be modulated either by a change in the "sensitivity" of postsynaptic receptors or in the number of transmitter packets released by an action potential, the quantal content. Much effort has been made to identify the site of induction and expression of long-term synaptic modulation (i.e., synaptic plasticity) as the first step for elucidating the underlying molecular mechanism. At the presynaptic terminal, a variety of G-protein-coupled receptors are expressed and contribute to presynaptic modulation. Among them, metabotropic glutamate receptors (mGluRs) and GABA$_B$ receptors are well known to suppress release of transmitters when activated by their ligands. Although the target of these presynaptic inhibitors were assumed to be either calcium channels, potassium channels, or the exocytotic machinery downstream of calcium influx, no direct evidence has been provided because of technical difficulties of recording from mammalian presynaptic terminals. A breakthrough in this situation was made recently by the development of a rodent brainstem slice preparation that has enabled us to record directly from presynaptic terminals with patch-clamp techniques.

The α_1-Subunit of the L-Type Ca^{2+} Channel Is Converted to a Long Open and Noninactivating State by Large Depolarization

S. NAKAYAMA[1], M. KUZUYA[2], M. MIYOSHI[1], K. KUBA[1], and Y. OKAMURA[3]

[1]Department of Physiology and [2]Department of Gerontology, School of Medicine Nagoya University, 65 Tsuruma-cho, Showa-ku, Nagoya 466-8550 Japan
[3]Biomolecular Engineering Department, National Institute of Bioscience and Human Technology, Higashi 1-1, Tsukuba, Ibaraki 305, Japan

Key words. Calcium channel, Multiple open states, Dihydropyridine, α_{1C}, Voltage-dependent conversion

Summary. Voltage-sensitive Ca^{2+} channels are distributed in a wide range of excitable cells, and are involved in numerous physiological functions, e.g., synaptic transmission in nerves and contractile activity in muscle. In the present study we investigated characteristics of voltage-dependent modulation of L-type Ca^{2+} channels, using cloned channel components. In all three expression systems tested (α_{1C-a} and α_{1C-b} in CHO cells, and α_{1C-a} in *Xenopus* oocytes), closure of the L-type Ca^{2+} channel was significantly slowed after large conditioning depolarizations. This delayed closure was obtained in the presence of Bay K 8644. These results suggest that the α_1-subunit of the L-type Ca^{2+} channel possesses at least two open states: the normal open state and a long open state, which is distinct from the so-called 'mode 2' gating. Also, coexpression of the β-subunit seems to modify the transition from the normal to the long open state induced by a large depolarization.

A standard cell-attached patch-clamp technique was applied to CHO (Chinese hamster ovary) cells in which the α_1-subunit cloned from rabbit lung (α_{1C-b}) is stably expressed (Welling et al. 1993; Hofmann and Klugbauer 1996). Bay K 8644 (2 μM) were contained in the patch pipette to increase the availability of the Ca^{2+} channels (Nakayama and Brading 1996), and 50 or 100 mM Ba^{2+} was used as a charge carrier. A test depolarizing step (+40 mV, 7.5 ms) was applied at fixed intervals, and every other test step was preceded by a large conditioning depolarization (+80 mV to +100 mV, 4 s).

After the large conditioning depolarization for 4 s, the cloned Ca^{2+} channels were still open and showed slow channel closure upon repolarization (to -60 mV). The sum of unitary currents revealed that the tail currents seen after large depolarizations had a slower deactivation time constant than without preconditioning depolarization. Essentially the same results were obtained in cardiac muscle α_1-subunits (α_{1C-a}: Mikami et al. 1989) expressed in CHO cells. These results suggest that the α_1-subunit of the L-type Ca^{2+} channel preserves a large depolarization-induced second open state reported in smooth muscle cells: in this open state Ca^{2+} channels do not, or only very slowly, inactivate during depolarization (Nakayama and Brading 1993b; Nakayama and Brading 1995b), and also deactivate slowly upon repolarization (Nakayama and Brading 1993a). Since the pipette contained Bay K 8644, the second open state seems distinct from so-called 'mode 2' gating (Nakayama and Brading 1995a). This voltage-dependent modulation (the second open state) of L-type Ca^{2+} channels, if it occurs in excitable cells, would potentiate Ca^{2+} entry upon the repolarizing phase of the action potential.

By the use of the long channel opening feature after large depolarization, the current-voltage relationship was measured directly. A ramp step (+80 to -80 mV, 40 ms) preceded by a conditioning depolarization (+80 mV, 4 s) was repeated. Subtraction between current traces with and without channel opening revealed that the slope conductance of the α_1-subunit of the L-type Ca^{2+} channel was ~25 pS, being consistent with the value previously estimated in this Ca^{2+} channel subunit without preconditioning depolarization (Bosse et al. 1992).

When the α_1-subunits described (α_{1C-a} or α_{1C-b}) were coexpressed with skeletal muscle β-subunit (β_{1a}), the observation of slow channel closure after a large conditioning depolarization of 4-s duration was significantly less frequent. The probability of the observation was recovered by reducing the duration of the preconditioning step (to 100-200 ms). The suppression of the transition of α_{1C} into the second open state may be due to the fast inactivating effect of the skeletal muscle β-subunits.

Also, cloned α_1-subunits (α_{1C-a}) were expressed in *Xenopus* oocytes, and Ca^{2+} channel currents were measured using a two-electrode voltage-clamp amplifier. The extracellular solution contained 20 mM Ba^{2+} and Bay K 8644 (2-10 μM). Cl⁻ was substituted with methanesulfonate, and 1 mM niflumic acid was added. In this expression system, too, slow deactivating tail currents were observed only after large conditioning depolarization. Similar results were produced even after a long preincubation (14 hrs) of H-7 (100 μM). Such a high concentration of H-7 would non-specifically inhibit cellular kinase activities. Thus, voltage-dependent phosphorylation mechanism does not seem to play a major role in the formation of the second open state. A possible explanation involves multiple energy levels in voltage sensors of L-type Ca^{2+} channels (Nakayama et al 1996).

The EM-induced inhibition of $I_{Ca(h)}$ was reversed completely by both treatment or pretreatment with 100 µM N-methylmaleimide, which is known as a sulfhydryl alkylating agent that inhibits G-protein action in NG108-15 cells (Kasai 1990). Pretreatment with 100 ng/ml pertussis toxin (PTx) for 18 h also completely eliminated the inhibition by EM1 and EM2. An identical effect of PTx on µ-opioid receptors in NGMO-251 cells that mediate inhibition of $I_{Ca(h)}$ by DAMGO has been reported previously (Morikawa et al. 1995). Together, these results indicate that the EM-induced inhibition of the Ca^{2+} channel current is mediated by the activation of the G_i/G_o type of G-proteins, which couples to the µ-opioid receptors. However, it is not clear at this moment which α or $\beta\gamma$ subunit of G proteins mediates this effect in NGMO-251 cells.

Little or no inhibition of $I_{Ca(h)}$ was induced by focal application of 100 nM EM1 and EM2 in the parental NG108-15 cells, expressing predominantly δ-receptors (Morikawa et al. 1995). However, substantial inhibition of 28.8 ± 4.72% (n=14) and 23.2 ± 3.94% (n=10) was obtained by application of 100 nM DPDPE in NG108-15 and NGMO-251 cells, respectively, reflecting the response mediated by endogenous δ-receptors that are present in both cell lines.

The results clearly show that the brain neuropeptides EM1 and EM2 caused marked inhibition of the high voltage-activated Ca^{2+} channel current after interacting with cloned µ-receptors expressed in NGMO-251 cells. Our results in transformed NG108-15 cells suggest that the novel peptides EM1 and EM2 function as neurotransmitters which modulate conductance of voltage-activated Ca^{2+} channels.

References

Higashida H, Hosh N, Knijnik R, et al (1998) Endomorphins inhibit high-threshold Ca^{2+} channel currents in rodent NG108-15 cells overexpressing µ-opioid receptors. J Physiol 507:71-75

Kasai H (1991) Tonic inhibition and rebound facilitation of a neuronal calcium channel by a GTP-binding protein. Proc Natl Acad Sci USA 88:8855-8859

Morikawa H, Fukuda K, Kato S, et al (1995) Coupling of the cloned µ-opioid receptor with the ω-conotoxin-sensitive Ca^{2+} current in NG108-15 cells. J Neurochem 65:1403-1406

Zadina JE, Hackler L, Ge L-J, et al (1997) A potent and selective endogenous agonist for the µ-opiate receptor. Nature 386:499-502

Introductory Review: Ca²⁺ Dynamics, and Modulation

K. KUBA

Department of Physiology, Nagoya University School of Medicine, 65 Tsurumai-chou, Showa-ku, Nagoya 466-8550,

Key words. Intracellular Ca^{2+}, Ca^{2+} pump, Na/Ca exchange, Ca^{2+} release, Ca^{2+}-induced Ca^{2+} release, Ryanodine receptor, IP_3 receptor, Voltage-Dependent Ca^{2+} channel, Glutamate receptor, Endoplasmic Reticulum, Mitochondria, Exocytosis, Ca^{2+} wave, Ca^{2+} microdomain, Modulation, Ca^{2+} oscillation

Summary. Free Ca^{2+} in the cytosol of neurones at rest is at quick equilibrium with a large amount of bound Ca and maintained at a low concentration, 1/10000 of external Ca^{2+} level, by extrusion at the cell membrane and uptake into Ca^{2+}-storing organelles. Exogenous or spontaneous cell activity increases cytosolic Ca^{2+} concentration ($[Ca^{2+}]_i$) via Ca^{2+} entry at the cell membrane and/or Ca^{2+} release from Ca^{2+} stores. Ca^{2+} release operates independently or amplifies a rise in $[Ca^{2+}]_i$ produced by Ca^{2+} entry or Ca^{2+} release (Ca^{2+}-induced Ca^{2+} release), causing the propagation of a high $[Ca^{2+}]_i$ wave in the cytosol and possibly in the nucleus. An increase in $[Ca^{2+}]_i$ directly or indirectly regulates cellular mechanisms as an intracellular messenger via activation of biochemical cascades; phosphorylation or dephosphorylation of proteins, activation of protein synthesis, and gene induction.

1 Physiological Actions of Cytosolic Ca²⁺ in Neurones

Regulatory actions of Ca^{2+} in neurones are more multimodal than those in other cells (Fig. 1). An increase in $[Ca^{2+}]_i$ activates the exocytosis of neurotransmitter (Katz 1969; Schweizer et al. 1995), modulates it in short- and long-term manners (Kuba and Kumamoto 1990; Zucker 1996; Milner et al. 1998), regulates Ca^{2+}-dependent K channels affecting the duration of an action potential (Robitaille and Charlton 1992), enhances the recycling of synaptic vesicles (Schweizer et al. 1995; Stevens and Wesseling 1998), and regulates the differentiation of neurones (Holliday et al. 1991). In postsynaptic neurones, cytosolic Ca^{2+} regulates Ca^{2+}-dependent K⁺ cation or Cl⁻ channels producing a slow afterhyperpolarization or depolarization (Kuba et al. 1983: see Chapter 3). Cytosolic Ca^{2+} plays important roles in the induction of long-term potentiation and depression via various ways such as changes in the sensitivity or availability of transmitter receptors or the activation of genes or protein synthesis involved in synaptic transmission (Milner et al. 1998). In mitochondria, $[Ca^{2+}]_i$ activates Ca^{2+}-dependent dehydogenases (Hansford 1985) and enhancing ATP synthesis (Hajnóczky et al. 1995).

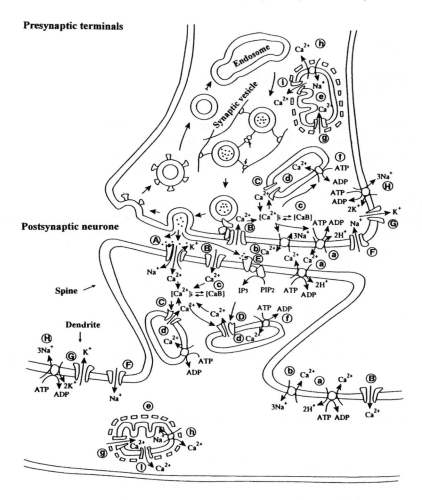

Fig. 1. A schema of Ca^{2+} dynamics in a synapse Ⓐ: receptor channel, Ⓑ: voltage-dependent Ca^{2+} channel, Ⓒ: ryanodine receptor, Ⓓ: IP_3 receptor, Ⓔ: transmitter receptor, Ⓕ: Na^+ channel, Ⓖ: K^+ channel, Ⓗ: Na^+/K^+ pump, ⓐ: Ca^{2+} pump, ⓑ: Na^+/Ca^{2+} exchanger, ⓒ: Ca^{2+}-binding protein, ⓓ: endoplasmic reticulum, ⓔ: mitochondria, ⓕ: Ca^{2+} pump, ⓖ: Ca^{2+} uniporter, ⓗ: Na^+/Ca^{2+} exchanger, ⓘ: membrane transition pore.

2 Maintenance of Low [Ca^{2+}]$_i$

Most Ca^{2+} ions in the cytoplasm are normally bound to Ca^{2+}-binding proteins and come into quick equilibrium with bound Ca in response to a change in [Ca^{2+}]$_i$ (Neher and Augustine 1992; Kasai 1993: Fig. 1). The ratio of bound Ca to free Ca^{2+} is around 100 (Neher and Augustine 1992). The low level of [Ca^{2+}]$_i$ is maintained by active extrusion of Ca^{2+} by two major Ca^{2+} transport systems at the cell membrane. One of them is Ca^{2+} pump, which operates as Ca-ATPase to normally pump one Ca^{2+} out of the cell in exchange for the uptake of two H$^+$ by consuming the energy of two molecules of ATP. This is known to operate at a relatively low level of [Ca^{2+}]$_i$, but to have a relatively low capacity so that it serves to maintain the resting level of [Ca^{2+}]$_i$ (Calafoli 1987). Another Ca^{2+} extrusion system is the Na/Ca exchanger, which normally extrudes one Ca^{2+} to the extracellular space in exchange for three Na$^+$, the downhill movement of which provides energy for Ca^{2+} extrusion (Calafoli 1987; Reuter and Porzig 1987: Fig. 1).

A large amount of Ca^{2+} is stored in Ca^{2+}-storing organelles. One major Ca^{2+} store is the endoplasmic reticulum (Blaustein 1988), which is the endowed with Ca^{2+} pumps and Ca^{2+} release channels at the membrane and Ca^{2+}-binding proteins in the lumen, enabling them to keep more Ca than they do as free Ca^{2+} (Meldolesi et al. 1992). The Ca^{2+} pump at the organelle membrane takes up two Ca^{2+} into the cytoplasm as Ca-ATPase by utilizing the energy of one ATP. Thus, Ca^{2+} uptake via Ca^{2+} pump and Ca^{2+} leakage through a small fraction of opened Ca^{2+} release channels are at equilibrium such that a Ca^{2+} concentration gradient exists that is one to ten thousand times higher in the lumen (Carafoli 1987). Another intracellular Ca^{2+} store is the mitochondrion, which does not normally retain Ca^{2+} as much as the endoplasmic reticulum does. When [Ca^{2+}]$_i$ level reaches more than several hundred nanomoles, it takes up Ca^{2+} strongly via the Ca^{2+} uniporter driven by a large electrochemical potential across the inner membrane. Once the [Ca^{2+}]$_i$ level has recovered to less than this level, Ca^{2+} is released from mitochondria via Na/Ca exchanger (Hansford 1985; Denton and McCormack 1990: Fig. 1). Thus, the mitochondria operate as a slow Ca^{2+} buffer, which dumps a rise in [Ca^{2+}]$_i$ caused rather by strong cell activity and prolongs the duration of the period of [Ca^{2+}]$_i$ increase (Friel and Tsien 1994; Nowicky and Duchen 1998). Other possible Ca^{2+} stores in neurones may be synaptic vesicles. This is based on the suggestion that chromaffin granules can store Ca^{2+} and release it (Petersen 1996) and also the observation of the existence of Ca^{2+} deposits in synaptic vesicles as revealed by X-ray microprobe analysis of preganglionic terminals of bullfrog sympathetic ganglia (Fujimoto et al. 1980).

It may be worthwhile to point out that the extracellular space is not infinite, but extremely restricted in small volume in situ. In this regard, it can be considered also as an important Ca^{2+} store in limited amount.

3 Pathways for $[Ca^{2+}]_i$ Increase

Two major pathways for Ca^{2+} entry at the neuronal membrane are voltage-dependent Ca^{2+} channels (VDCC) and transmitter-operated receptor channels(Fig. 1). The type and characteristics of VDCC are reviewed in the preceding chapter. Glutamate receptors of NMDA type (Sommer and Seeburg 1992) and a subtype of AMPA type (Ozawa et al. 1991; Hollmann and Heinemann 1994; Ozawa et al. 1998 for review) and nicotinic acetylcholine receptors are endowed with an ion channel that is highly permeable to Ca^{2+} in addition to Na^+ and K^+. The NMDA receptor is normally blocked by Mg^{2+} at resting membrane potential and relieved from the blockade by membrane depolarization. This provides a voltage-dependent regulation of Ca^{2+} entry, which is important for the induction of long-term potentiation (LTP) of transmission in central synapses by conjoined activation of NMDA- and non-NMDA-type glutamate receptors (Malenka and Nicoll 1993). The key molecular structure for high Ca^{2+} permeability of NMDA- and Ca^{2+}-permeable AMPA-type glutamate receptors is a site of the M_2 domain of gluR2 subunit that contains aspartate and glutamin, respectively. This position is occupied by arginin in non-Ca^{2+}-permeable AMPA-type glutamate receptors (Sommer and Seeburg 1992). Ca^{2+}-permeable AMPA-type glutamate receptors are suggested to play roles in Ca^{2+}-dependent disorders of neurones as well as other Ca^{2+}-dependent functions (Pellegrini-Giampietro et al. 1997). Nicotinic acetylcholine receptors are involved in fast synaptic transmission of some types of central synapses (Frazier et al. 1988) as well as peripheral synapses and presynaptic modulation (Radcliffe and Dani 1988). The reversed mode of operation of the Na/Ca exchanger, which is known in cardiac muscles, does not seem to be reported for neurones. Ca^{2+} channels activated by depletion of Ca^{2+} stores (Putney 1991), which is known in nonneuronal cells, do not appear to be found in neurones.

There are two types of Ca^{2+} release channels. The ryanodine receptor is predominantly activated by cytosolic Ca^{2+}. This mode of Ca^{2+} release is called Ca^{2+}-induced Ca^{2+} release (CICR; Kuba 1994). The activation of CICR results in the amplification of a rise in $[Ca^{2+}]_i$ (Ca^{2+} signal) caused by Ca^{2+} entry or other Ca^{2+} release mechanisms. The activation of ryanodine receptors is enhanced by the endogenous modulator, cyclic ADP-ribose, produced from βNAD (Lee et al. 1989; Hua et al. 1994). Unfortunately, physiological stimuli that produce cyclic ADP-ribose have not been found yet. Inositoltrisphosphate (IP_3) receptors are activated by IP_3 produced from phosphatidylinositol or other cascades in response to the activation of metabotropic receptors (Berridge, 1993). The sensitivity of IP_3 receptors to IP_3 is strongly dependent on $[Ca^{2+}]_i$. A moderate rise in $[Ca^{2+}]_i$ increases the sensitivity, while a large rise in $[Ca^{2+}]_i$ suppresses it (Bezprozvanny et al. 1991). Thus, a rise in $[Ca^{2+}]_i$ itself activates IP_3 receptors at a subthreshold level of intracellular IP_3 concentration, causing CICR. This mode of activation appears to be a predominant mechanism of activation of IP_3 receptor in some types of neuronal functions.

Mitochondrion has two Ca^{2+} release pathways. One is the Na/Ca exchanger, which operates in a rather passive mode, when the intramitochondrial Ca^{2+} level is high in relative to $[Ca^{2+}]_i$ (Friel and Tsien 1994). Another Ca^{2+} release mechanism from mitochondria is a permeability transition pore, which is a large nonselective ion channel opened by a rise in intramitochondrial pH (Bernardi et al. 1993; Szabó and Zoratti 1993). A rise in $[Ca^{2+}]_i$ results in Ca^{2+} entry into mitochondria, increases luminal pH, and then opens the channels, releasing Ca^{2+} (Ichas et al. 1997). Thus, this indicates that CICR indeed occurs at the mitochondria.

4 Ca^{2+} Microdomain

When VDCC at the cell membrane or Ca^{2+} release channels at the organelle membrane are opened, a high $[Ca^{2+}]_i$ region appears at the orifice of channels in the cytoplasmic side (Fogelson and Zucker 1985; Simon and Llinas 1985; Llinas et al. 1992). This high $[Ca^{2+}]_i$ region, called Ca^{2+} microdomain, dissipates in space and time initially by diffusion and binding to Ca^{2+}-binding proteins and then by localized Ca^{2+} uptake and extrusion. The temporal and spatial dynamics of the Ca^{2+} mircrodomain were estimated by computer simulation (Fogelson and Zucker, 1985; Simon and Llinas 1985). The $[Ca^{2+}]_i$ in the Ca^{2+} microdomain reaches several hundred micromoles in a fraction of a millisecond in a hemisphere of a diameter of several tens of nanometers and quickly decreases in concentration, but broadens in space. Within several tens of milliseconds the increased $[Ca^{2+}]_i$ decreases to several hundred nanomoles, but become almost homogenous if it occurs in a small restricted structure, such as presynaptic terminals, spines, and dendrites. If Ca^{2+} channels or Ca^{2+} release channels exist in clusters, they would form a larger Ca^{2+} "macrodomain." For instance, in a dendritic tree of Purkinje fibers a localized jump of IP_3 concentration produced a high $[Ca^{2+}]_i$ domain restricted in a spine or small regions of dendrites (Finch and Augustine 1998; Takechi et al. 1998).

The high $[Ca^{2+}]_i$ microdomain in a short period would favor the fast action of Ca^{2+} via a low affinity Ca^{2+}-binding protein, which activates the fast physiological function such as the exocytosis of neurotransmitter (Fig. 1). On the other hand, the more global but moderate rise in $[Ca^{2+}]_i$ would conform to the rather slow and sustained actions of Ca^{2+} via high-affinity Ca^{2+}-binding proteins. Thus, the localized existence of the origins of Ca^{2+}, localized or global, strong Ca^{2+}-buffering system and the different sensitivity of Ca^{2+}-binding proteins provide the differential actions of Ca^{2+} in time and space in neurones.

1996), it is also possible that a global increase in $[Ca^{2+}]_i$ is caused by intracellular Ca^{2+} release from Ca^{2+} stores (Kuba and Nishi 1976; Kuba 1980; Thayer et al. 1988; Lipscombe et al. 1988; Llano et al. 1994; see review by Kuba 1994). In this article, we show evidence for the activation of Ca^{2+}-induced Ca^{2+} release (CICR) in response to tetanic Ca^{2+} entry in frog motor nerve terminals and its unique characteristics and discuss the physiological significance.

2 Methods

An intracellular recording technique was applied to frog sartorius muscles to record miniature end-plate potentials (MEPPs) as a measure of relative changes in $[Ca^{2+}]_i$ in motor nerve terminals. Confocal measurement of fluorescence of Ca^{2+}-sensitive probes, Indo-1 or Oregon Green BAPTA-1, loaded in nerve terminals was made in frog cutaneus pectoris muscles. The ratio of the intensities of Indo-1 fluorescence measured at two wavelength ranges or the ratio of the fluorescence intensity of Oregon Green BAPTA-1 during or after stimulation to that before stimulation were converted to $[Ca^{2+}]_i$ values (Narita et al. 1998). In most experiments, we have used low Ca^{2+} (0.05-0.2 mM), high Mg^{2+} (10 mM) solutions, in which impulse-induced rises in $[Ca^{2+}]_i$ are so small that no end-plate potentials (EPPs) occur. Under this condition, the contribution of CICR to a total rise in $[Ca^{2+}]_i$ produced by tetanic stimulation would be greater, if it occurs.

3 Results and Discussion

3.1 Transient Rises in MEPP Frequency and $[Ca^{2+}]_i$ in Nerve Terminals During Continuous Tetanus

Continuous tetanic stimulation at 50 Hz applied to the nerve produced an increase in $[Ca^{2+}]_i$ in nerve terminals of frog cutaneus pectoris muscles, which grew slowly in onset, then regeneratively to the peak, and decayed to a level higher than the pretetanic level (Ca^{2+} hump: Figs. 1A, 2B). A similar transient rise in MEPP frequency during continuous tetanus was seen in frog sartorius muscles (MEPP hump: Fig. 2A). The MEPP hump depended on the extracellular Ca^{2+} concentration ($[Ca^{2+}]_o$) and stimulation frequency. Decreasing $[Ca^{2+}]_o$ from 0.2 to 0.05 mM almost abolished MEPP humps. Decreasing stimulation frequency from 50 to 20 Hz reduced the peak amplitude of MEPP hump and slowed its rate rise. Both Ca^{2+} and MEPP humps were abolished by blockers of ryanodine receptor, ryanodine (10 μM: Sutko et al. 1979) and TMB-8 (8-(N,N-diethylamino)octyl-3,4,5-trimethoxybenzoate hydrochloride: 4-10 μM: Fig. 3 for Ca^{2+} hump) (Chiou and Malagodi 1975) and a blocker of Ca^{2+} uptake into Ca^{2+}-storing organella,

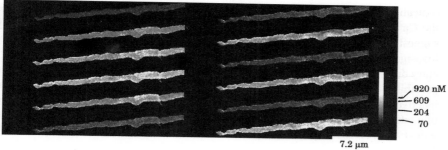

920 nM
609
204
70

7.2 µm

Fig. 1A,B. Tetanus-induced increases in $[Ca^{2+}]_i$ in frog motor nerve terminals. A 50 Hz tetanus was given to the nerve throughout the experiments. All the images were recorded in a low Ca^{2+}, high Mg^{2+} solution and are shown by the ratio of the fluorescence of Oregon Green BAPTA-1 to that before tetanus. **A** The image in the top is the control, which is the ratio of an image to another before the beginning of stimulation. The second to sixth images are those taken at 1, 2, 3, 4 and 5 min after the beginning of tetanus. The ratio values averaged over each image are used in the initial part of the graph in Fig. 2A. **B** Increases in $[Ca^{2+}]_i$ in the motor nerve terminals after pauses of different duration during 50 Hz tetanus. All the images are shown in ratios to that before the beginning of tetanus. The first, third, and fifth images from the top were taken at the end of each pause, while the second, fourth, and sixth images were taken at the peak of increases after pauses of 10, 30, and 60 s. The images were taken immediately after recording the images shown in A from the same cell. The ratio values averaged over each image fulfilled the data points in the later part of the graph in Fig. 2A.

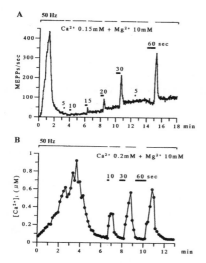

Fig. 2A,B. Ca^{2+} and MEPP humps and the dependence of postpause rises in $[Ca^{2+}]_i$ and MEPP frequency on pause duration. All the records were obtained in low-Ca^{2+}, high-Mg^{2+} solutions. **A** MEPP hump and rises in MEPP frequency after brief pauses of different duration in tetanus (50 Hz) applied following a MEPP hump. **B** A Ca^{2+} hump and increases in $[Ca^{2+}]_i$ after brief pause in tetanus (50 Hz). The data points were plotted from the experiments shown in Fig. 1.

thapsigargin (1 µM), but enhanced by a mitochondrial poison, CN (2 mM). Thus, the characteristics of the Ca^{2+} hump seen in nerve terminals of cutaneus pectoris muscles and those of the MEPP hump observed in sartorius muscles are similar, indicating that the MEPP hump is expected to occur in the former muscles, while the Ca^{2+} hump should take place in the latter muscles. The findings accordingly suggest that the Ca^{2+} hump is produced by the activation of CICR from thapsigargin-sensitive Ca^{2+} stores, but not from mitochondria, in response to Ca^{2+} entry during tetanus and that the resultant rise in $[Ca^{2+}]_i$ causes a MEPP hump. The possibility that the decay phase of the Ca^{2+} hump results from the Ca^{2+}-dependent inactivation of voltage-dependent Ca^{2+} channels, leading to a transient appearance of a tetanus-induced rise in $[Ca^{2+}]_i$, was easily ruled out by the strong quenching of indo-1 fluorescence by Mn^{2+} applied externally after a Ca^{2+} hump (Narita et al. 1998).

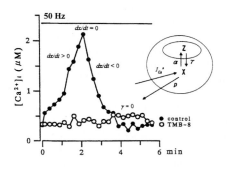

Fig. 3. Effects of TMB-8 on Ca^{2+} hump and a schema for Ca^{2+} dynamics in frog motor nerve terminals. All the recordings were made from Indo-1-loaded terminals in a low Ca^{2+} (0.2 mM), high-Mg^{2+} (10 mM) solution. The ratio (F_{412}/F_{475}) of fluorescence peaking at 412 nm (F_{412}) to that at 475 nm (F_{475}) was converted to $[Ca^{2+}]_i$ values. Effects of TMB-8 (10 μM) were seen 30 min after treatment with the drug, which began following a period of no stimuli for 30 min. *Inset* is a schema to illustrate Ca^{2+} dynamics involving CICR in frog motor nerve terminals. Ca^{2+} enters into the motor nerve terminals through voltage-dependent Ca^{2+} channels $(J_{Ca}{}^*)$, and is extruded out of the terminal by Ca^{2+} pumps (p) at the cell membrane. Cytoplasmic Ca^{2+} (x) activates Ca^{2+} release (γ) from Ca^{2+} stores and is taken into Ca^{2+} stores by Ca^{2+} pumps (α) to resume Ca^{2+} concentration in Ca^{2+} stores (z).

3.2 The Logical Basis for the Understanding of Ca^{2+} Dynamics During the Activation of CICR

We have used a simple model (inset in Fig. 3) in which basic elements for intracellular Ca^{2+} homeostasis of a neurone and those for rises in $[Ca^{2+}]_i$ (x) are assumed to exist in frog motor nerve terminals. Ca^{2+} release (rate constant; γ) occurs in response to Ca^{2+} entry from Ca^{2+} stores (z) via ryanodine receptors, and Ca^{2+} released is taken up by thapsigargin-sensitive Ca^{2+} stores (rate constant; α) and extruded out of the cell by a mechanism (rate constant; p) including Ca^{2+} pumps and Na/Ca exchange at the cell membrane. Changes in the rate of cytosolic Ca^{2+}, dx/dt, is expressed by the sum of the rates of Ca^{2+} release, γz and Ca^{2+} entry, $J_{Ca}{}^*$, and the rates of Ca^{2+} uptake, αx, and extrusion, px. Thus, an equation similar to that previously proposed for CICR (Kuba and Takeshita 1981) is derived with the inclusion of Ca^{2+} buffering by Ca^{2+}-binding proteins (Neher and Augustine

1992) and mitochondria $(1 + \kappa_s)$ (Duchen et al. 1990; Thayer and Miller 1990; Friel and Tsien 1994).

$$(1 + \kappa_s)\, dx/dt = J_{Ca}^* + \gamma z - (\alpha + p)x \qquad \text{or} \qquad (1)$$

$$\gamma z = (1 + \kappa_s)dx/dt + (\alpha + p)x - J_{Ca}^* \qquad (2)$$

In the equation, Ca^{2+} influx into Ca^{2+} stores through Ca^{2+} release channels can be implicitly ignored.

To analyze the kinetic behavior of CICR based on this model, we have to know what sorts of experimental parameters reflects the rate of Ca^{2+} release. Ca^{2+} entry produced by each action potential during tetanus is considered to be constant, since Ca^{2+}-dependent inactivation of voltage-dependent Ca^{2+} channels (Eckert and Tillotson 1981) appears to be small, as suggested from the quick quenching effect of Mn^{2+} on the fluorescence of Ca^{2+} probes. Furthermore, J_{Ca}^* in low Ca^{2+}, high Mg^{2+} solutions is negligibly small when compared with the amount of Ca^{2+} released. Thus, Ca^{2+} entry may be omitted from interpretation of Ca^{2+} dynamics during and after tetanus . Then, Equation 2 implies that the sum of $(1 + \kappa_s)dx/dt$ and $(\alpha + p)x$ would reflect the rates of Ca^{2+} release (γz).

When $(\alpha + p)x$ is much greater than $(1 + \kappa_s)dx/dt$ (or dx/dt is negligibly small), it becomes

$$\gamma z \approx (\alpha + p)x \qquad (3)$$

This condition is found to hold for the time course of a Ca^{2+} -hump (see Narita et al., 1998 for calculation).

On the other hand, when $(\alpha + p)x$ is smaller than $(1 + \kappa_s)dx/dt$

$$\gamma z \approx (1 + \kappa_s)dx/dt \qquad (4)$$

When the values of dx/dt are measured at similar values of x, changes in dx/dt would reflect changes in the rate of Ca^{2+} release. This can be measured by applying short pauses in tetanus, which result in a quick decrease in $[Ca^{2+}]_i$ during the course of a Ca^{2+} hump (Fig. 4A). Then, changes in the initial rate of rise in $[Ca^{2+}]_i$ measured immediately after the resumption of tetanus would yield alterations in γz during the course of a Ca^{2+} hump (Fig. 4B). The time course of alterations in γz during the course of a Ca^{2+} hump is similar to that of the Ca^{2+} hump.

3.3 The Priming, Activation, Inactivation, Deinactivation, and Depriming of the CICR Mechanism

The time course of CICR in response to repetitive Ca^{2+} entries can be reflected in the time course of a component of a Ca^{2+} hump blocked by TMB-8 (Fig. 3). This demonstrates that CICR occurs from the very beginning of tetanus, although it is initially very slowly, and disappears rather quickly in spite of the existence of trigger Ca^{2+} due to Ca^{2+} entry.

When $dx/dt > 0$, namely during the rising phase of Ca^{2+} hump, the rate of Ca^{2+} release exceeds the sum of the rates of Ca^{2+} uptake and extrusion. Under this condition, the amount of Ca^{2+} (z) in Ca^{2+} stores must be decreasing. This idea and the fact that the activation of CICR occurs from the very beginning of tetanic stimulation suggest that Ca^{2+} stores are normally filled with Ca^{2+} before tetanus, one of the important conclusions in the understanding of Ca^{2+} dynamics of CICR. The rising phase of a Ca^{2+} hump itself and an increase in dx/dt measured immediately after the resumption of a pause during that phase (see the preceding section) indicate an increase in γz during the rising phase of a Ca^{2+} hump (Fig. 4A). Because z is decreasing during this phase, γ must slowly increase during this phase (Fig. 4B). This process, a slow elevation of the probability of Ca^{2+} release or the conversion to Ca^{2+} release channels from not activable state to an activable state, may be called "priming".

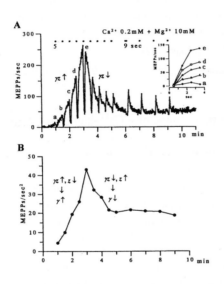

Fig. 4A,B. The time course of CICR activation: effects of brief pauses in tetanus on MEPP frequency during and after a MEPP hump. **A** Changes in MEPP frequency caused by a pause in tetanus (50 Hz) applied during and after a MEPP hump. Pauses of 5 or 9 s were applied during a period indicated by a horizontal bar. *Inset* is the expansions of the rising phases of MEPP frequency after a short pause (5 s) given at 1 (**a**), 1.5 (**b**), 2 (**c**), 2.5 (**d**), and 3 min (**e**) after the beginning of a tetanus (50 Hz). The abscissa is the time after resumption of tetanus, while the ordinate represents a net increase in MEPP frequency. **B** Changes in the initial rate of rise of MEPP frequency after a pause during the course of a series of tetani (50 Hz). The rate of rise of MEPP frequency was measured from the slope of the regression line fitted to the initial three data points. Note that the time for the peak of the rate of rise roughly corresponds to the peak of MEPP hump shown in A.

When $dx/xt < 0$, namely during the falling phase of the Ca^{2+} hump, the sum of the rates of Ca^{2+} uptake and extrusion exceed the rate of Ca^{2+} release. Furthermore, the falling phase itself and the decrease in dx/dt during that phase indicate a decrease in γz under this condition (Fig. 4A), although the latter parameter was not reduced to zero because of the effects of the pause to remove the inactivation of Ca^{2+} release channels (see below). It is likely that the decrease in γz can mostly be ascribed to a decrease in γ (Fig. 4B), but not to a reduction in z, although it is not known how great a fraction of Ca^{2+} released is taken up into Ca^{2+} stores or extruded out of the terminals. There is evidence that a fairly large amount of Ca^{2+}

is stored during and after the falling phase of a Ca^{2+} hump. When a short pause was applied during and after a Ca^{2+} hump, there was a sharp rise in $[Ca^{2+}]_i$ after the resumption of tetanus, which was equivalent in amplitude to that of a Ca^{2+} hump (Fig. 2A). Furthermore, there was no change in the level of $[Ca^{2+}]_i$ during the pause applied after the end of a Ca^{2+} hump (Fig. 2A). This suggests that a fairly large amount of Ca^{2+} remained or was taken up during and after the falling phase of a Ca^{2+} hump. In addition, it is obvious that the release of Ca^{2+} stops at the end of a Ca^{2+} hump in spite of the continuation of repetitive Ca^{2+} entry. Accordingly, the rate constant for Ca^{2+} release (γ) decreases during the falling phase of a Ca^{2+} hump.

The mechanism of the decrease in γ may be explained by the removal of the "primed state" of Ca^{2+} release channels or the inactivation of the channels. The latter mechanism is likely for the following reason. The postpause rises in $[Ca^{2+}]_i$ and MEPP frequency (indirectly reflecting changes in the level of $[Ca^{2+}]_i$) produced after the end of Ca^{2+} and MEPP humps was augmented as the duration of pause was increased (Fig. 2A,B). This is opposite to the priming effect of tetanus on the CICR mechanism. Accordingly, the dependence of magnitude of the postpause rise on pause duration can be explained by the deinactivation of the CICR mechanism. The depriming process was seen by a decrease in the initial rate of the postpause rise produced by a longer pause (>1 min).

The time course of activation of CICR in frog motor nerve terminals is summarized as follows. Ryanodine receptor/Ca^{2+} release channels are slowly primed in response to repetitive Ca^{2+} entries during tetanus. Once Ca^{2+} release channels are primed, they are swiftly activated by subsequent Ca^{2+} entry. Further continuation of repetitive Ca^{2+} entry inactivates the Ca^{2+} release channels. The Ca^{2+} entry-dependent inactivation is quickly removed by stopping Ca^{2+} entry for a short period (<1 min). A longer cessation (>1 min) of Ca^{2+} entry resulted in the depriming of Ca^{2+} release channels over tens of minutes.

3.4 The Physiological Significance of the CICR Mechanism in Presynaptic Terminals

The existence of CICR via ryanodine receptors in frog motor nerve terminals found in this study is consistent with previous studies. Caffeine and theophylline, activators of ryanodine receptor, enhanced both the spontaneous and impulse-evoked release of transmitter in several types of presynaptic terminals (Elmqvist and Feldman 1965; Onodera 1973; Ohta and Kuba 1980; Tóth et al. 1990). Peng (1996) and Smith and Cunnane (1996) reported some evidence for the activation of CICR in autonomic presynaptic terminals and involvement in transmitter release.

The MEPP hump observed in sartorius muscles and the Ca^{2+} hump seen in cutaneus pectoris muscles are similar in time course, the mode of responses to pauses in tetanus and pharmacology. Thus, both MEPP and Ca^{2+} humps can be interpreted on the same ground. Accordingly, a rise in $[Ca^{2+}]_i$ by the activation of et al. unpublished data; Deisseroth et al. 1996). As already mentioned, generalized action potential firing, which should largely favor opening of N-type, P/Q-type, and T-type Ca^{2+} channels in the dendrites (Kavalali et al. 1997), was not sufficient in generating a visible nuclear phosphorylation event (Deisseroth et al. 1996). This was further supported by the finding that blockers of the N-type and P/Q-type channels (ω-CTx-GVIA and ω-Aga-IVA) had no effect on either CREB phosphorylation or CaM translocation (Deisseroth et al. 1998). In contrast, activation of NMDA receptor or L-type Ca^{2+} channels in isolation was sufficient in inducing nuclear pCREB formation (Bito et al. unpublished data; Deisseroth et al. 1998). Thus a selective coupling was found between NMDA receptor/L-type Ca^{2+} channel activation and the CREB signaling system. The molecular mechanism underlying such specificity is currently under investigation.

4 Conclusion

We and others have shown the multiplicity and the diversity of dendritic Ca^{2+} entry sources in hippocampal neurons. Activation of synaptic glutamate receptors can initiate opening of NMDA receptors and various types of voltage-gated channels when sufficient summation of EPSPs takes place. Alternatively, some channels such as N- or P/Q-type Ca^{2+} channels could conversely be inhibited by glutamate via heterotrimeric G-proteins. Such a differential regulation of distinct sources of Ca^{2+} mobilization may play a role in shaping dendritic excitability and local Ca^{2+} signaling. It also may contribute, at least in part, to the apparently privileged role for L-type Ca^{2+} currents in the synaptic activity-driven CREB phosphorylation. Furthermore, localized synaptic Ca^{2+} signals via a subset of Ca^{2+} influx channels, in the absence of a global Ca^{2+} increase, may be more amenable to subtler modulations, such as those required to achieve the dual Ca^{2+}-dependent control of CREB phosphorylation and dephosphorylation, and thus may provide a basis for the efficient coupling of CREB phosphorylation to CREB-dependent gene transcription (Bito et al. 1997) (Fig.2).

Acknowledgments. This work was supported in part by grants from the Japan Brain Foundation, the Yamanouchi Foundation for Research on Metabolic Disorders and a Sasakawa Scientific Research Grant from the Japan Science Society (to H.B.), an American Heart Association (California Affiliate) postdoctoral fellowship (to E.T.K.), a Medical Scientist Training Program fellowship (to K.D.), and an NIMH-Silvio Conte Center Grant (to R.W.T.).

References

Bading H, Ginty DD, Greenberg ME (1993) Regulation of gene expression in hippocampal neurons by distinct calcium signaling pathways. Science 260:181-186

Bean B (1985) Two kinds of calcium channels in canine atrial cells: differences in kinetics, selectivity and pharmacology. J Gen Physiol. 86:1-30

Bito H (1998) The role of calcium in activity-dependent neuronal gene regulation. Cell Calcium 23:143-150

Bito H, Deisseroth K, Tsien RW (1996) CREB phosphorylation and dephosphorylation: a Ca^{2+}- and stimulus duration-dependent switch for hippocampal gene expression. Cell 87:1203-1214

Bito H, Deisseroth K, Tsien RW (1997) Ca^{2+}-dependent regulation in neuronal gene expression. Curr Opin Neurobiol 7:419-429.

Deisseroth K, Bito H, Tsien RW (1996) Signaling from synapse to nucleus: postsynaptic CREB phosphorylation during multiple forms of hippocampal synaptic plasticity. Neuron 16:89-101

Deisseroth K, Heist EK, Tsien RW (1998) Translocation of calmodulin to the nucleus supports CREB phosphorylation in hippocampal neurons. Nature 392:198-202.

Ginty DD, Kornhauser JM, Thompson MA,et al (1993) Regulation of CREB phosphorylation in the suprachiasmatic nucleus by light and a circadian clock. Science 260:238-241

Hille B (1994) Modulation of ion-channel function by G-protein-coupled receptors. Trends Neurosci 17:531-536

Kay AR, Wong RKS (1986) Isolation of neurons suitable for patch-clamping from adult mammalian central nervous systems. J Neurosci Methods 16:227-238

Kavalali ET, Zhuo M, Bito H, et al (1997) Dendritic Ca^{2+}-channels characterized by recordings from isolated hippocampal dendritic segments. Neuron 18:651-663

Llinas R, Sugimori M (1980) Electrophysiological properties of *in vitro* Purkinje cell dendrites in mammalian cerebellar slices. J Physiol 305:197-213

Magee JC, Johnston D (1995) Synaptic activation of voltage-gated channels in the dendrites of hippocampal pyramidal neurons. Science 268:301-304

Markram H, Sakmann B (1994) Calcium transients in dendrites of neocortical neurons evoked by single subthreshold excitatory postsynaptic potentials via low-voltage-activated calcium channels. Proc Natl Acad Sci USA 91:5207-5211

Markram H, Helm PJ, Sakmann B (1995) Dendritic calcium transients evoked by single back-propagating action potentials in neocortical pyramidal neurons. J. Physiol. 485:1-20

Nilius B, Hess P, Lansman JB et al. (1985) A novel type of cardiac calcium channel in ventricular cells. Nature 316:443-446

Randall A, Tsien RW (1995) Pharmacological dissection of multiple types of Ca^{2+} channel currents in rat cerebellar granule neurons. J Neurosci 15:2995-3012

Spruston N, Schiller Y, Stuart G, et al (1995) Activity-dependent action potential invasion and calcium influx into hippocampal CA1 dendrites. Science 268:297-300

Wheeler DB, Randall A, Tsien RW (1994) Roles of N-type and Q-type Ca^{2+} channels in supporting hippocampal synaptic transmission. Science 264:107-111

Wong RKS, Prince DA, Basbaum AI (1979) Intradendritic recordings from hippocampal neurons. Proc. Natl. Acad. Sci. USA 76:986-990

Yuste R, Gutnick MJ, Saar D, et al (1994) Ca^{2+} accumulations in dendrites of neocortical pyramidal neurons: an apical band and evidence for two functional compartments. Neuron 13:23-43 images were collected at 0.5- to 1-s intervals (excitation, 365 ± 30 nm bandpass; dichroic, 480 nm; emission, 510 nm) using a high-speed cooled CCD camera (Hamamatsu Photonics model 4880-80, Hamamatsu, Japan) and a Pentium computer (Digital) with the HiSCa software (Hamamatsu Photonics, Hamamatsu). Drugs (10 µl) were pressure applied (100 ms) with a glass pipette (tip 6-8 µm in diameter) located about 70 µm upstream from the recording field.

2.3 Ca $^{2+}$ Imaging

For Ca^{2+} imaging in CHO cells, the dishes were incubated for 30 min at $37°C$ in BSS solution containing 3 µM Fura-2 AM (Molecular Probes, Eugene, OR, USA). They were then allowed to stay for 30 min in BSS prior to the experiment. The cells were placed on an inverted microscope and observed as described for thermal imaging except that they were alternatively illuminated at 350 and 380 nm and observed at 510 nm.

2.4 Simultaneous Ca^{2+} and PKC Imaging

For simultaneous Ca^{2+} and PKC imaging, PC12 cells were incubated for 30 min at $32°C$ in Krebs medium (106 mM NaCl, 4.5 mM KCl, 1.2 mM $MgSO_4$, 2.5 mM $CaCl_2$, 11 mM D-glucose, 1.2 mM KH_2PO_4, 25 mM $NaHCO_3$, pH 7.4) containing 2 µM Fura-2-AM. The cells were then washed with Krebs medium and subsequently incubated for 30 min with 0.2 µM Fim-1 in Krebs solution. They were then placed on an inverted microscope (Axiovert 35M Zeiss, Germany). They were superfused with Krebs solution (1 ml/min) and drugs were applied by superfusion for 30 s. The cells were alternatively illuminated at 350 nm (for Fura-2 fluorescence) and at 490 nm (for Fim-1 fluorescence). For both probes the emission was observed with a long-pass filter at 510 nm. Images were recorded every 10 s for each excitation wavelength using an intensified CCD camera (Extended Isis, Photonic Science, UK) and the Fluostar software (Imstar, Paris, France).

2.5 Image Analysis

In every case the image series was analyzed using the same procedure. After background subtraction from the experimental image series, a virtual images series was calculated by exponential interpolation to assess for the fluorescence baseline during the whole recording period. To normalize the response, a ratio image series was then calculated by dividing experimental images by interpolated baseline images. Regions of interest were defined on the ratio images and finally semiquantified. This process was either automatically done by the Hamamatsu

HiSCa setup or performed on a DecAlpha workstation (Digital Co, Boston, MA, USA) using software developed in the laboratory.

2.6 Immunohistochemistry

Differentiated PC12 cells were treated for immunolabeling with antibodies raised against the PKCα isoform (rabbit polyclonal anti-PKCα, Sigma, St Louis, MO, USA). Cultured cells were fixed with 0.2% glutaraldehyde in phosphate-buffered saline (PBS, pH 7.4) for 20 min. The fixation was followed by a 15- to 30-min reduction step (0.5% sodium borohydride in PBS) and membrane permeabilization for 20 min in methanol-PBS (1:4 vol/vol). The cells were then incubated for 20 min in 10% normal goat serum in PBS to block nonspecific binding and for an additional 2-h period with the anti-PKCα and anti-PKCδ antibodies (1/800 dilution, in PBS containing 5% normal goat serum) at room temperature. A goat antirabbit secondary antibody (1/200 in 5% goat serum) fluorescein isothiocyanate-conjugated (Jackson) was applied for 1 h. Cultures were finally mounted in Vectashield™ medium (Vector) to reduce photobleaching. Controls were obtained by omitting the primary antibody in the incubation bath, and in these conditions no staining of the cells was observed (not shown).

3 Results

3.1 Metabotropic Glutamate Receptor Followed by Thermal and Ca²⁺ Imaging

3.1.1 Effects of Agonists and Antagonists

Eu-TTA detects heat production by decreasing its fluorescence while heat absorption results in a signal increase. We applied three different agonists of mGluR1 receptors to the cultured cells. 10 μM glutamate or 30 μM S-3,5-dihydrophenylglycine (DHPG; Tocris, Bristol, UK) application evoked a large heat absorption in the cells (Figure 1A and E, respectively). The response evoked by 100 μM (1S,3R)-1-aminocyclopentane-1,3-dicarboxylic acid (t-ACPD; Tocris, Bristol, UK) was similar although with a smaller amplitude than with the other agonists (Figure 1B). In each case the signal sharply increased after a delay phase (20 to 30 s), reaching a slightly decreasing plateau and then decreased towards the baseline. The response was dose-dependent because 10 μM DHPG evoked a weak response while 100 μM DHPG application was saturating (not shown).

reduced. The same effect was also observed when the cells were loaded with Fura-2 (Figure 2E,F).

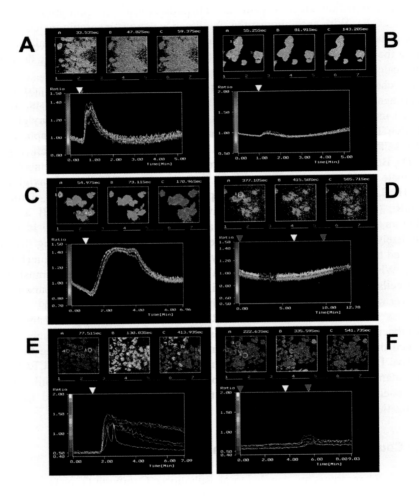

Fig. 2A-F. A, B 30 µM DHPG-induced response in control cells (A) and in Pertussis toxin (PTX) pretreated cells (B). The response is mostly inhibited by PTX treatment. **C-F** Effect of 100 nM thapsigargin superfusion (*red arrowheads* in C and F) on 10-µM glutamate response. Thermal imaging (C,D) and calcium imaging (E,F) of the glutamate-induced response in the absence (C, E) and in the presence (D, F) of thapsigargin.

3.2 Muscarinic Receptor Activation Followed by Simultaneous Ca²⁺ and PKC Activation Imaging

To investigate the cellular response downstream of calcium mobilization in the living cells, we studied the effect of carbachol application on PC12 cells by loading the cells with Fura-2 and Fim-1. This procedure allowed us to simultaneously record intracellular calcium changes and the activation of PKC. Adding increasing carbachol concentrations in the perfusion medium leads to the activation of muscarinic receptors, which are expressed in differentiated PC12 cells. As expected we observed dose-dependent intracellular calcium changes in the stimulated cells (Figure 3A). These changes were observed together with increases in Fim-1 fluorescence intensity (Figure 3A). In a previous study we have shown that Fim-1 fluorescence increase indicates PKC activation (de Barry et al. 1997), a process which presumes PKC translocation from a soluble form to a membrane-bound form. This induces a conformational change, which unmasks the enzymatic catalytic site and allows the fluorescent probe to interact with the protein. In PC12 cells the perfusion of GÖ 7874, a specific PKC inhibitor, almost completely abolished Fim-1 signal when 100 μM carbachol was applied to the cells while the Fura-2 response was unaffected (not shown).

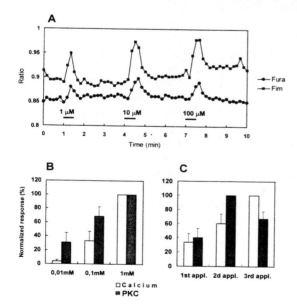

Fig. 3A-C. A Simultaneous recording of calcium and PKC responses induced by increasing carbachol concentrations applied on PC12 cells. The cells were previously loaded with Fura-2 and Fim-1 and the agonist was superfused for 30 s every 3 min. **B** Histogram of the dose-response relationship of the carbachol effect on these cells. **C** Desensitization of PKC response induced by three sequential applications of 1 mM carbachol for 30 s. Note the partial PKC response desensitization while the calcium response is amplified.

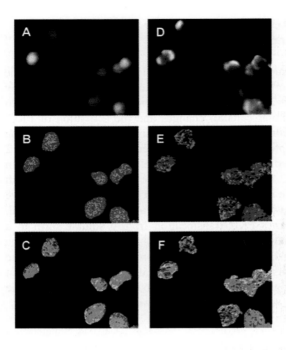

Fig. 4A-F. PC12 cells simultaneously loaded with 2 µM Fura-2 and 0.2 µM Fim-1. **A** Fura-2 fluorescence image observed when the cells were illuminated at 350 nm. **B, C** Ratio images of Fura-2 fluorescence when the preparation was unstimulated (B) or in the presence of 100 µM carbachol (C). **D**: Fim-1 fluores-cence of the same field when the cells were excited at 490 nm. **E, F** Fim-1 ratio images of the cells at rest (E) and stimulated by carbachol (F). Note the high PKC activation levels in discrete sites along the plasma membranes.

Upon 30-s 100-µM or 1-mM carbachol application, the response increased within 15 s and started to decrease immediately after the stimulus withdrawal to reach the baseline within 1 min (Figure 3A). However, although the kinetics of the two fluorescent signals appeared similar, the desensitization of Fim-1 response was rapid as compared to the Fura-2 signal when 1 mM carbachol was applied three times consecutively for 30 s (Figure 3B). A careful examination of the images showed that the response to KCl application did not occur at the same location inside the cells. The calcium rise affected the whole cytoplasm inside the cell, whereas PKC activation mainly occurred in patches at the periphery of the cell, presumably close to the plasma membrane (Figure 4). Actually an immunocytochemical staining of differentiated PC12 cells revealed such a membrane patch distribution for PKC α and δ isoforms (Figure 5).

Discussion

4.1 Thermal Imaging

As shown by using different agonists and antagonists, the observed signal is dose-dependent and is inhibited by group I mGluR antagonists. Hence, it appears to be induced by mGluR1 activation because other mGluRs types are not expressed in these cells. The weak heat production observed at the beginning of the stimulation sequence during the latency phase and the additionnal heat absorption observed at the end of the sequence are correlated with the association and with the dissociation of the ligand, respectively. This result suggests that these weak signals may reflect ligand binding to the receptor protein. Interestingly, however, these signals were observed neither in the presence of an antagonist nor after PTX treatment, which means they would be due to the conformation change of the receptor or even the dissociation/association of the G-protein.

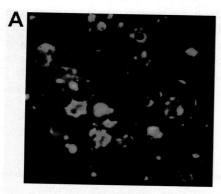

Fig. 5A,B. Immunocytochemistry of differentiated PC12 cells labelled with antbodies directed against PKCα (A) and PKCδ (B) isoforms. Note the heterogeneous distribution of the label at the periphery of the cell.

The question was then to know which cellular process was responsible for the large heat absorption observed during the cellular response. PTX only inhibits G_i- and G_o-proteins by ADP ribosylation. In our case PTX was able to inhibit the response. We could thus say that probably a G_o protein was involved in the signal transduction process. Moreover, we could also conclude that heat absorption reflected an event downstream from G-protein activation in the cellular metabolic cascade. The thapsigargin effect showed us that the observed heat absorption was mainly correlated with intracellular Ca^{2+} mobilization or an event downstream from calcium mobilization. But since the thapsigargin inhibition of the thermal signal and of the Ca^{2+} response was not complete, we could not exclude the possibility that part of heat absorption was due to phospholipase C (PLC) activity and inositide hydrolysis. In this respect, it is interesting to point out that in some cells using DHPG as mGluR1 agonist we could detect a plateau phase with two maxima. The first maximum was mostly observed at the periphery of the cells, while the second maximum was related to the cytosolic response of the cells. The first maximum occurred a few seconds after agonist application, and considering its location it may reflect a membrane event such as PLC activity. Its onset, however, is rather slow (a few seconds) when compared to the calcium response (Figure 2C,E). Another explanation would thus be that the observed heat absorption is not a primary effect of cell stimulation but more likely a secondary effect from a rapid recovery process inside the cell. These hypotheses are possible because of time and space resolutions given by thermal imaging only. This technique, although not quantitative, appears complementary to and easier to perform than microcalorimetry.

4.2 PKC Imaging

The PC12 cell line from rat pheochromocytoma express a neuronal phenotype when they are differentiated: they grow processes, are excitable, express muscarinic receptors, and can secrete catecholanime when stimulated. Indeed they express several PKC isoforms including α and β, which are Ca^{2+} dependent, and δ, ε, and ζ, which are Ca^{2+} independent. PKC is a soluble protein when inactivated. When the protein interacts with phosphatidylserine (PS) and diacyglycerol (DAG) in the plasma membrane, its conformation changes, unmasking its catalytic site and allowing enzymatic activity. This model was thus used to study cellular events, which are triggered by DAG synthesis and calcium mobilization. Carbachol applications on PC12 cells induced a dose-dependent response of intracellular Ca^{2+} and of PKC activation. The kinetics of the two responses were similar, indicating a close correlation between the two phenomena. Surprisingly, the PKC response went rapidly back to the baseline, indicating a reversible activation of the enzyme. This result is apparently in contradiction to biochemical measurements in various models, which have shown a persistent PKC activation after stimulation. However these assays were performed by isolating membrane-bound PKC and adding phorbol ester, a nonphysiological PKC activator. According to our measurements, the Ca^{2+} trigger of PKC activation is extremely sensitive, and thus difficult to control in biochemical assays, and the phorbol ester activation is only partially reversible. We suggest that PKC, while still binding to the membrane or interacting with membrane proteins like AKAP-79 or RACKS, may undergo a reversible activation triggered by DAG and/or intracellular calcium. The heterogeneous distribution of the PKC activation sites in the cell as shown in our study and their preferential location in the cellular membrane would strengthen this hypothesis. The rapid desensitization of the PKC response upon sequential stimulation also illustrates the changes in the cellular response, which may lead to plasticity in neuronal cells. In vitro experiments (unpublished results) suggest that this desensitization is due to a partial hydrolysis of a specific activable PKC pool by Ca^{2+} dependent proteases such as μ-calpain. Since these processes are rather rapid, it is noteworthy to point out that these results could not be revealed by classical biochemical studies lacking time and space resolution, but only an imaging method.

5 Conclusions

These imaging techniques are suitable to study ligand-receptor interactions as well as the cellular metabolic cascades in various preparations. They allow investigations of fast biochemical cellular processes. They also show the cellular compartmentation of the metabolic cascades, which might change our concepts about these phenomena. Indeed numerous methods are currently being developed for such a purpose; many of them use recombinant or conjugated fluorescent proteins. The advantages of using small permeant molecules as in the present case are simpler experimental conditions, often a rapid and good diffusion of the probes, and no steric hindrance.

Acknowledgments. This work was supported by grants from the "Research for the Future" program (RFTF) 96L00310 from the Japanese Society for the Promotion of Science (JSPS) to T. Y. and from the ACC-SV12 program from the French ministry for Education and Research to J.B.

References

de Barry J, Kawahara S, Takamura K, et al (1997) Time resolved protein kinase C imaging in sea urchin egg during fertilization. Exp Cell Res 234:115-124

Crosby GA, Whan RE, Alire RM (1961) Intramolecular energy transfer in rare earth chelates. Role of the triplet state. J Chem Phys 34:743-748

Winston H, Marsh OJ, Suzuki CK, et al (1963) Fluorescence of europium theonyltrifluoroacetone. I. Evaluation of laser threshold parameters. J Chem Phys 39:267-271

Zohar O, Ikeda M, Shinagawa H, et al. (1998) Thermal imaging of receptor-activated heat production in single cells. Biophys J 74:82-89

3 Results

3.1 Glu Receptor-Mediated Ca^{2+} Mobilization in the SCN

In Mg^{2+} free-ACSF, Glu (10 μM-1 mM), AMPA (1-50 μM), and NMDA (1-100 μM) dose dependently increased intracellular Ca^{2+} in the hypothalamic area. The largest Ca^{2+} response was induced by NMDA and the spatial distribution of Ca^{2+} rise was particularly dense inside the SCN. The NMDA-induced Ca^{2+} response was significantly reduced both by competitive and noncompetitive NMDA-antagonists, DL-APV (100 μM) and MK-801 (100 μM), respectively, demonstrating the receptor-mediated Ca^{2+} entry. The amplitude of the NMDA (100 μM)-induced response was not dependent on the circadian time of observation in Mg^{2+}-free ACSF. However, in normal ACSF containing 1 mM Mg^{2+}, the response was significantly larger during the daytime than during the nighttime. It is known that the SCN slice displays a higher action potential firing frequency during the daytime, and the potency of the Mg^{2+} block of the NMDA receptor depends on the membrane potential. Therefore, these data suggest that the Ca^{2+} permeability of NMDA receptor, which is widely distributed in the SCN, is regulated by the endogenous rhythms of membrane potential of the SCN neurons.

3.2 GABA-A Receptor-Mediated Ca^{2+} Mobilization in the SCN

The daytime (11:00-17:00) application of GABA (1 mM) or the GABA-A agonist, muscimol (30 or 50 μM) induced Ca^{2+} mobilization in the SCN, while nighttime (23:00-5:00) application induced smaller or no responses. In addition, the muscimol increased the $[Ca^{2+}]i$ in presence of TTX (1 μM), but the response was significantly reduced by simultaneous application of D-APV (50 μM) (Fig. 1A). Therefore, GABA-A-mediated Ca^{2+} mobilzation may be caused by the depolarization of the postsynaptic membrane and the resultant removal of the Mg^{2+} block of the NMDA receptors, although there are also considerable voltage-dependent Ca^{2+} channels that could contribute to the Ca^{2+} mobilization.

GABA-A-mediated Ca^{2+} mobilization is known to occur in hippocampus neurons at early developmental stages (Leinekugel et al. 1997). We also observed a developmental change of GABAergic Ca^{2+} responses in the SCN neurons. Indeed, the Ca^{2+} responses caused by GABA (1 mM) and muscimol (50 μM) were larger in young neonatal animals (PD 6-9). However, the GABA-A-mediated Ca^{2+} response was sustained during PD 11-20 (Fig.1B). Therefore, these data demonstrate that stimulation of GABA-A receptors can trigger Ca^{2+} mobilization via NMDA receptors even in the adult SCN.

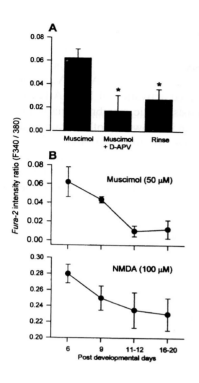

Fig. 1 A N-methyl-D-aspartate (NMDA) antagonist, D-APV significantly blocked muscimol-induced Ca^{2+} mobilization in the suprachiasmatic nucleus (SCN) of 10 days old mice. *$P<0.05$ in comparison with the response by muscimol alone. Therefore, GABA-A-mediated Ca^{2+} mobilization may be due to the activation of NMDA receptors, which are densely distributed in the SCN. **B** Developmental changes in the Ca^{2+} mobilization in the SCN induced by muscimol and NMDA. Note that muscimol-induced Ca^{2+} responses were sustained even at the adult stages, although both the NMDA- and muscimol-induced Ca^{2+} responses were reduced in the course of development. All these experiments were examined during the daytime since the Ca^{2+} responses were also circadian time dependent (see details in Results).

3.3 Cl$^-$ Imaging of SCN

The mechanism of GABA-A-mediated depolarization is still controversial but one proposed mechanism involves the efflux of Cl$^-$ (Luhmann and Prince, 1991; Rohrbough and Spitzer, 1996). To determine whether or not such inverse Cl$^-$ flux occurs in the SCN, we used MEQ-based Cl$^-$ imaging techniques. Muscimol (50 μM) application onto the SCN slice at the time which muscimol increases Ca^{2+} (PD 10, during the daytime) induced an immediate decrement of [Cl$^-$]$_i$ followed by the increment of [Cl$^-$]$_i$. Therefore, initial Cl$^-$ efflux through the GABA-A receptor channels may induce the excitatory actions of GABA-A in the SCN.

4 Discussion

The present Ca^{2+} imaging analysis revealed that stimulation of GABA-A as well as NMDA receptors increased Ca^{2+} throughout the SCN, although the amplitude of GABA-mediated mobilization of Ca^{2+} was smaller than the NMDA-mediated response. These Ca^{2+} mobilizations were similarly larger during the daytime and

Enhancement of Neurotransmitter Release by Activation of Ryanodine Receptors after Ca^{2+}-Dependent Priming at Motor Nerve Terminals

K. NARITA[1], T. AKITA[2], K. OCHI [1] and K. KUBA[2]

[1]Department of Physiology, Kawasaki Medical School, 577 Matsushima, Kurashiki 701-0192
[2]Department of Physiology, School of Medicine, Nagoya University, 65 Tsurumai-cho, Showa-ku, Nagoya 466-8550, Japan

Key words: Presynaptic terminal, Ryanodine receptor, Ca^{2+}-dependent priming , Ca^{2+}-induced Ca^{2+} release, Transmitter release

Summary. Ryanodine receptors at frog motor nerve terminals need priming by repetitive Ca^{2+} entry before activation by further Ca^{2+} entry. Ryanodine receptors were primed by Ca^{2+} entry at a rate of >2 Hz at the external Ca^{2+} concentration > 0.05 mM, while their activation required Ca^{2+} entry at a rate of >20 Hz. Ca^{2+}-induced Ca^{2+} release via ryanodine receptors thus primed and activated markedly augmented impulse-induced release of transmitter from frog motor nerve terminals over several tens of minutes.

1 Introduction

In frog motor nerve terminals, repetitive Ca^{2+} entry slowly primes, then activates, ryanodine receptors and causes Ca^{2+}-induced Ca^{2+} release (CICR) from thapsigargin-sensitive and CN-insensitive Ca^{2+} stores, leading to a marked enhancement of asynchronous release of transmitters (Narita et al. 1998). Two questions were addressed: how the priming and activation of CICR depend on the frequency of Ca^{2+} entry and extracelluar Ca^{2+} concentration ([Ca^{2+}]$_o$), and how the activation of CICR affects impulse-induced transmitter release.

2 Methods

End-plate potentials (Epps) and miniature Epps (Mepps) were recorded from end-plates of frogs (*Rana nigromaculata*) by a conventional intracellular recording technique. Intracellular free Ca^{2+} concentration ($[Ca^{2+}]_i$) in the nerve terminals was recorded by measuring changes in the fluorescence of the K salt of Indo-1 or Oregon Green BAPTA-1 with a confocal laser scanning microscope (Biorad, MRC-600) or cooled CCD camera (Hamamatsu Photonics, HisCa) with an image analysis software (Hamamatus Photonics, Argus-50). Ca^{2+} probes were loaded from the cut end of the motor nerve overnight (Narita et al. 1998).

3 Results and Discussion

Priming and activation of CICR in frog motor nerve terminals can be seen as slow and subsequent regenerative transient rises in $[Ca^{2+}]_i$ and Mepp frequency (Ca^{2+}- and Mepp humps, respectively) during continuous tetanus of 50 Hz in a low Ca^{2+} (0.2 mM), high Mg^{2+}(10 mM) solution. Ca^{2+}- and Mepp humps occurred over a few minutes and were suppressed by blockers of ryanodine receptors, TMB-8 and ryanodine. Ca^{2+}- and Mepp humps were abolished by lowering $[Ca^{2+}]_o$ to 0.05 mM, while Ca^{2+}- and Mepp humps similar in onset and time course to those at 0.2 mM $[Ca^{2+}]_o$ and 10 mM $[Mg^{2+}]_o$, but in a greater magnitude, were induced in normal Ringer containing d-tubocurarine. Thus, priming of CICR occurs at $[Ca^{2+}]_o$ greater than 0.05 mM and becomes maximum at 0.2 mM in the presence of 10 mM $[Mg^{2+}]_o$ (Narita et al. 1998).

0.5 mM Ca²⁺, 10 mM Mg²⁺

Control

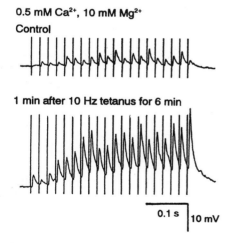

1 min after 10 Hz tetanus for 6 min

0.1 s   10 mV

Fig. 1. Epps induced at 50 Hz before and after application of a conditioning tetanus 10 Hz for 6 min.

The dependence of the priming process on stimulation frequency differs from that of activation. A tetanus at 20 Hz produced only a small and slow regenerative rise in Mepp frequency, indicating that conspicuous activation of CICR needs repetitive Ca^{2+} entry, at least at an interval <50 ms. On the other hand, priming occurs at much lower frequency. The frequency dependence of the priming process was examined by combination of a tetanus varying in a low frequency range (0.5~10 Hz) for priming of CICR and a tetanus of 50 Hz for activation. Following a low frequency tetanus >2 Hz, a high frequency tetanus produced a fast regenerative rise in Mepp frequency. Thus, priming takes place by repetitive Ca^{2+} entry at an interval <0.5 ms and grows with increases in its rate. Priming was also dependent on the duration of repetivite Ca^{2+} entry.

The activation of CICR markedly augments impulse-evoked release of transmitter. After the application of a priming tetanus of 10 Hz for 10 min in a low Ca^{2+}(0.5 mM), high Mg^{2+}(10 mM) solution, Epps induced by a 50-Hz tetanus were markedly enhanced in amplitude and rate of growth during the tetanus. The enhancement was depressed by ryanodine and TMB-8 and subsided over tens of minutes as a result of depriming of the CICR mechanism.

The results suggest that priming of the CICR mechanism at frog motor nerve terminals occurs by the slow binding of Ca^{2+} to a molecule of a relatively high affinity involved in CICR, while activation takes place by the fast binding of Ca^{2+} to a relatively high-affinity site of a molecule involved in CICR.

Reference

Narita K, Akita T, Osanai M, et al (1998). A Ca^{2+}-induced Ca^{2+} release mechanism involved in asynchronous exocytosis at frog motor nerve terminals. J Gen Physiol 112:593-609

Upregulation of Cytosolic Ca^{2+} Increases by Cyclic ADP-ribose in NG108-15 Neuronal Cells: In Comparison with Inositol Tetrakisphosphate in Fibroblast Cells

M. HASHII and H. HIGASHIDA

Department of Biophysical Genetics, Kanazawa University Graduate School of Medicine, 13-1 Takara-machi, Kanazawa 920-8640, Japan

Key words. Cyclic ADP-ribose, β-NAD$^+$, Ryanodine receptors, Ins(1,4,5)P$_3$, L-type voltage-activated Ca^{2+} channels, NG108-15 cells

It has been reported that the depolarization-induced increase in [Ca^{2+}]$_i$ is potentiated by cyclic ADP-ribose (cADPR) or caffeine in neurons (Hua et al. 1994; Kano et al. 1995). Modulation of L-type voltage-activated Ca^{2+} channels (VACCs) by caffeine and ryanodine has also been shown in cerebellar neurons (Chavis et al. 1996). However, mechanisms whereby ryanodine receptors, cADPR, and L-type VACCs interact have not been established yet. Thus, we examined the effects of cADPR on a [Ca^{2+}]$_i$ rise in NG108-15 cells. NG108-15 cells were pre-loaded with fura-2 and whole-cell patch-clamped (Hashii et al. 1996). Application of 10 μM cADPR itself did not trigger any [Ca^{2+}]$_i$ rise at the resting membrane potential of -45 mV. However, cADPR augmented the increased [Ca^{2+}]$_i$ resulting from Ca^{2+} mobilization. Extracellular applications of bradykinin (20 μl of 2.5 μM, final 100 nM) to NG108-15 cells evoked an immediate and transient [Ca^{2+}]$_i$ rise in the presence of 2 mM [Ca^{2+}]$_o$. The [Ca^{2+}]$_i$ level at the peak and 3 min after the bradykinin application was about 280% and 110% of the concentration (73 nM) just before bradykinin applications. Usually, after cADPR infusion the peak bradykinin response was significantly larger (4-fold) than the control (2.8-fold). The [Ca^{2+}]$_i$ level after 3 min of bradykinin application returned to the control level, at which time the effect of cADPR disappeared. Intracellular applications of 10 μM Ins(1,4,5)P$_3$ into NG108-15 cells elicited a transient [Ca^{2+}]$_i$ rise by 1.7-fold at the peak of the preinjection level. When Ins(1,4,5)P$_3$ and cADPR were simultaneously infused, the peak [Ca^{2+}]$_i$ was significantly increased to 2.2-fold of the control. The higher [Ca^{2+}]$_i$ level was maintained longer by cADPR. This result suggests that cADPR can further

Plastic Nature of a Ca²⁺-Induced Ca²⁺ Release Mechanism in Hippocampal Synaptic Terminals

F.-M. LU and K. KUBA

Department of Physiology, Nagoya University, School of Medicine, Nagoya 466-8550, Japan.

Key words. Transmitter release, Ca²⁺-induced Ca²⁺ release, High K⁺-induced exocytosis, Whole-cell patch clamp, Hippocampal neurones

Summary. In cultured autaptic hippocampal neurones patch-clamped with an electrode containing CsCl as a major salt, a rapid application of a moderately high K⁺ (3.75~110 mM) solution caused a transient outward current during, or inward current after, holding the membrane at +30 mV. These currents (K10 responses) were blocked by glutamate or GABA receptor antagonist, Co²⁺, Cd²⁺, and the removal of external Ca²⁺, but not by caffeine or ryanodine, greater when the interval of high-K⁺ treatments was longer, and summated when induced in succession. K-responses appear to be thus produced by intracellular Ca²⁺ release via Ca²⁺ release channels other than ryanodine receptor in presynaptic terminals depolarized by a K⁺-jump under the action of intracellular Cs⁺.

1 Introduction

Ca²⁺ entry into presynaptic terminals activates exocytosis of neurotransmitter. It is unknown, however, whether intracellular Ca²⁺ release causes exocytosis from presynaptic terminals except for peripheral synapses. We report here a high K⁺-induced release of transmitters from autaptic terminals of cultured rat hippocampal neurons, whose characteristics can not be explained solely by Ca²⁺ entry, but in part by Ca²⁺-induced Ca²⁺ release (CICR) differing from those involving ryanodine receptors.

2 Methods

Hippocampal neurons were cultured from E20-day-old Wistar rat embryos. Whole-cell patch clamp was made to the cell soma. Patch pipettes were filled with (in mM): CsCl, 132; MgCl$_2$, 2; EGTA, 10; HEPES (pH 7.2), 10; Na$_2$ATP, 2. Normal Krebs solution contained (in mM): NaCl, 137.8; KCl, 2.5; CaCl$_2$, 3; MgCl$_2$, 1; HEPES/NaOH (pH 7.3), 10; glucose, 25. Tetrodotoxin (0.5 μM) was added to all the test solutions. High K$^+$ solutions (3.75-15 mM) were applied through a polyethylene tube (inner diameter, 300 μm) placed at 0.8-1.0 mm from the cell soma by a rapid laminal flow changer.

3 Results and Discussion

All the experiments were done on neurons forming autapses. A rapid jump of external K$^+$ concentration ([K$^+$]$_o$) from 2.5 to 10 mM caused only a small inward current at holding potential (V$_H$) of –60 mV. When V$_H$ was maintained at + 30 mV, a rapid [K$^+$]$_o$ jump to 10 mM produced an outward current that consisted of a spike and subsequent plateau and gradually increased in amplitude. After repolarization to V$_H$ of –60 mV, a 10 mM [K$^+$]$_o$ jump induced an inward current (K10 response) with a steep rise and slow decay, which slowly decreased in amplitude with appearance and increase of a delay up to 6 s. K10 responses were blocked by either picrotoxin or a combination of CNQX and CPP, indicating that K10 responses are caused by the release of GABA or glutamate from autaptic terminals. Characteristics of K10 responses are (1) the least effective [K$^+$]$_o$ of 3.75mM, (2) summation of the responses by stepwise increases in [K$^+$]$_o$, (3) the longer the preceding interval, the greater and longer the K10 responses, (4) disappearance by the removal of external Ca^{2+} or addition of Cd^{2+} or Co^{2+}, and (5) the lack of effects of removal of external Ca^{2+} on the effect of conditioning cell membrane depolarization to produce K10 responses.

There are several possible mechanisms. First, the conditioning depolarizing current from the cell soma depolarizes the cell membrane of autaptic terminals and facilitates the diffusion of Cs$^+$ into the terminals, which would further depolarize the terminal membrane by blocking K$^+$ channels. The depolarization produced by 10 mM [K$^+$]$_o$ under this condition may be strong enough to activate voltage-dependent Ca^{2+} channels at the terminal membrane, which may not be effectively clamped at a V$_H$ equivalent to that at the cell soma. The resultant rise in intracellular Ca^{2+} ([Ca^{2+}]$_i$) would release neurotransmitters. Differences in the probability of Ca^{2+}-spike generation and in distance from the cell soma among terminals could cause variability of K10 responses. Gradual growth and dissipation of Cs$^+$ in the terminals by the holding currents at +30 and –60 mV, respectively, would explain the plastic nature of K10 responses. Second, somatic depolarizing current may prime the mechanism of CICR, while the current for holding V$_H$ at -60 mV would deprime the mechanism. Under this condition, Ca^{2+} entry induced by 10 mM [K$^+$]$_o$ via voltage-dependent Ca^{2+} channels and/or K$^+$-activated Ca^{2+} channels (Deák et al. 1998) would activate CICR and so the release of neurotransmitters.

SNAP-25 (Banerjee et al. 1996), does not affect docking or priming of synaptic vesicles. These results indicate that SNAREs do not play an essential role in docking.

A recent in vitro experiment revealed that SNAREpins (synaptobrevin-syntaxin complexes) are minimal machinery for membrane fusion (Weber et al. 1998). Weber et al. (1998) demonstrated that recombinant synaptobrevin and syntaxin associated with different lipid bilayer membranes forming separate vesicles assemble into SNAREpins at a low-energy stable state. This leads to spontaneous fusion of docked membranes at physiological temperature. SNAREpins are formed by coiled-coil interactions of α-helix of syntaxin, SNAP-25, and synaptobrevin (Sutton et al. 1998: Fig. 3). Strain produced in the α-helix of synaptobrevin and syntaxin embedded in the synaptic vesicle and plasma membranes, respectively, may be transduced to SNAREpins via extended α-helix into the pins. Resultant changes in charge distribution at the membrane-anchored end of fusion complex may promote vesicle fusion (Sutton et al. 1998).

According to this mechanism of fusion, disassembly of SNAREpins by an ATPase, NSF, together with SNAPs may thus occur after fusion of synaptic vesicles. In support of this, using syntaxin and NSF mutants of *Drosophila*, Littleton et al. (1998) found evidence that NSF disassembles SNAREs after fusion, when the SNAREs reside in the presynaptic membrane. They also suggested that endocytosis occurs independent of the SNARE complex. On the other hand, Xu et al. (1998) suggested that there is an equilibrium between assembly and disassembly of SNAREpins without exocytosis, based on the blockade of exocytosis by botulinum toxins in chromaffin cells, which are known only to act to SNARE proteins in disassembled conditions. Accordingly, it remains to be asked when and how SNARE disassembly occurs and, in particular, what specific state of the NFS and SNAPs-sensitive ternary complex corresponds to the progression of vesicle traffic.

3.2 Proteins Interacting with SNAREs

SNAREs are controlled by a multitude of other protein. Munc18 (n-Sec1), complexin and tomosyn are cytosolic proteins that associate with SNAREs and thought to regulate rhe SNARE complex formation. Detailed in vitro binding studies showed that Munc18 interacts with syntaxin that prevents binding of SNAP-25 or synaptobrevin and thereby precludes the formation of SNARE complex (Pevsner at al. 1994), while complexin and tomosyn promote SNARE complex assembly, respectively (McMahon et al. 1995; Fujita et al. 1998). Regulatory role of complexin in neurotransmitter release has also been suggested by functional studies at *Aplysia* buccal ganglia (Ono et al. 1998). Injection of recombinant complexin and α-SNAP into presynaptic neurons caused depression and facilitation of neurotransmitter, respectively. The effect of complexin was reversed by a subsequent injection of recombinant α-SNAP, and vice versa.

Augustine and his colleagues, however, demonstrated that peptides, a part of sequence of Munc18 or complexin, inhibited exocytosis in squid giant synapses, indicating that interaction of Munc18 or complexin with syntaxin is essential for docked vesicles to fuse (Dresbach et al. 1998; Tokumaru et al. 1997)(Fig. 3). SNARE complex interacts with N-type and P/Q type Ca^{2+} channels that provide Ca^{2+} for triggering exocytosis in the peripheral and central nervous system (Sheng et al. 1994; Rettig et al. 1996: Fig. 3). This interaction is found to be essential for synchronous neurotransmitter release, since a peptide of the syntaxin-binding site of N-type Ca^{2+} channel, blocked transmitter release at nicotinic synapses of rat superior cervical ganglion cells (Mochida et al. 1996). Snapin is a protein exclusively located on synaptic vesicle membranes and associates with the SNARE complex through direct interaction with SNAP-25. Binding of the recombinant C-terminal half of snapin injected into presynaptic neurons reversibly inhibited neurotransmitter release in rat superior cervical ganglion synapses (Ilardi et al. 1999).

3.3 GTP-Binding Proteins and Associated Proteins

A GTP-binding protein, Rab3, is a vesicle-associated protein having a GTPase motif and GTP/GDP binding domains (Touchot et al. 1987; Matsui et al. 1988). Rab3 and its binding proteins, rabphillin3A (Shirataki et al. 1993) and Rim (Wang et al. 1997) are involved in exocytosis via hydrolysis of GTP (Bean and Scheller, 1997: Fig. 3). Rab3A appears to operate in limiting the fusion machinery to cause exocytosis of a single synaptic vesicle in response to a single impulse (Geppert et al. 1997). On the other hand, Rim possessing zinc-finger motif and C2 domain is a protein associated with the plasma membrane at the active zone and binds to GTP-complexed Rab3A. Rim is suggested to serve as a regulator of synaptic vesicle fusion by forming a GTP-dependent complex between synaptic vesicles and plasma membranes (Wang et al. 1997). Castillo et al. (1997) demonstrated that rab3A is essential for mossy fiber long-term potentiation in the hippocampus.

3.4 Calcium-Binding Proteins

Calcium-binding proteins containing two C2 domains that interact with Ca^{2+} have been considered to act as Ca^{2+} sensors in nerve terminals. Synaptotagmin, a synaptic vesicle protein, which binds to Ca^{2+} and phospholipid with its C2 domains, has been most characterized as a Ca^{2+} sensor in exocytosis (Südhof 1995)(Fig. 3). Twelve isoforms of synaptotagmin have been identified. Synaptotagmin I (and II) directly interacts with syntaxin. This interaction is regulated by Ca^{2+} but requires more than 200 µM for half-maximal binding (Li et al. 1994; Chapman et al. 1995). This approximates the Ca^{2+} requirement of synaptic vesicle exocytosis and suggests a mechanism whereby Ca^{2+} triggers exocytosis by regulating synaptotagmin I (and II) interaction with syntaxin and

2 Synapses Formed Between SCGNS In Culture

These synapses have been employed in investigations of synapse formation and trophic factors. Isolated SCGNs form synapses with some other ganglionic neurons when cultured in the presence of nerve growth factor (Rees and Bunge 1974; Johnson et al. 1976). The neurite endings form presynaptic varicosities (Ko et al. 1976; Wakshull et al. 1979) that contain small clear synaptic vesicles (Johnson et al. 1976), and these synapses generate postsynaptic responses sensitive to nicotinic receptor blockers (O'Lague et al. 1974; Ko et al. 1976; Wakshull et al. 1979). A variety of externally applied factors and conditions lead to a reduction of synaptic catecholaminergic properties and simultaneously induce cholinergic properties in SCGNs in long-term culture (Landis 1990). Several cholinergic switching factors that might be released by the target tissue have been purified (Weber 1981; Fukuda 1985; Wong and Kessler 1987; Adler et al. 1989; Saadat et al. 1989; Rao and Landis 1990).

2.1 Synapse Preparation

Superior cervical ganglia were dissected from 7-day postnatal rats, desheathed, and incubated with collagenase (0.5 mg/ml; Worthington Biochem. Co.) in L-15 medium (Gibco) at 37°C for about 20 min. Following enzyme treatment, small tissue chunks were triturated gently through a small-pore glass pipette until a cloudy suspension was obtained. After washing by low-speed centrifugation (1200 rpm for 3 min) and resuspension, the collected cells were plated onto cover slips in plastic dishes (35 mm in diameter, approximately one ganglion per dish) containing the growth medium: 84% Eagle's minimum essential medium (Gibco), 10% fetal calf serum (MAB), 5% horse serum (Gibco), 1% penicillin/streptomycin solution (Gibco), and 25 ng/ml nerve growth factor (2.5 S, Collaborative Research). Cells were maintained at 37°C in a water-saturated atmosphere of 95% air-5% CO_2; the medium was changed twice per week. Somata appeared to be round (10-20 μm) when isolated flattened and processes extended and ramified over the cover slip dish within 1 week. The somata enlarged (30-50 μm) and neurites formed complex connections in 2-5 weeks of culture (Mochida et al. 1994a).

2.2 Presynaptic Terminal Proteins Expressed in SCGNs Synapses

Presence of the proteins identified in mature presynaptic terminals such as synaptic vesicle proteins as well as presynaptic plasma membrane proteins in these synapses formed in culture could be confirmed by immunofluorescence staining, indicating that these cultured synapses express many proteins characteristic of mature synapses (Mochida, 1995). Synaptophysin, a protein associated with synaptic vesicles (Jahn et al. 1985; Wiedenmann and Franke

1985), was found as small spots both around the soma and on the hillocks of processes of a solitary cell, and on the surface and inner aspect of neuron clusters. These spots of synaptophysin, presumably in synaptic vesicles, indicate that synaptic vesicles are present in presynaptic terminals making synapses with somata or hillocks of processes (Figure 1A,B). Spots of synaptophysin increased with time in culture. Presence of synaptotagmin (Matthew et al. 1981) and VAMP/synaptobrevin-2 (Elferink et al. 1989; Südhof et al. 1989) was detected. Proteins associated with presynaptic terminal plasma membrane, syntaxin (Bennett et al. 1992) and neurexin (Petrenko et al. 1991), were also expressed in these synapses. Antibody against syntaxin (1A and 1B) showed a similar distribution of syntaxin to that of VAMP or other synaptic vesicle-associated proteins, indicating that syntaxin molecules are concentrated in presynaptic terminals.

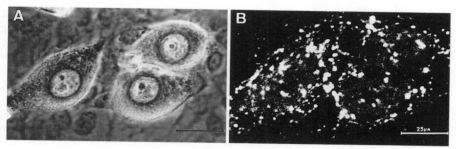

Fig. 1A-B. **A** Phase contrast image of superior cervical ganglion neurons (SCGNs) in culture. **B** Immunofluorescence staining image for synaptophysin using a monoclonal antibody with Texas Red-labeling (Boeringer). SCGNs were isolated from 7-day-old rats and were cultured for 31 days. Bar 25 μm.

2.3 Electrophysiological Recordings of Synaptic Responses

Conventional intracellular recordings are useful for measuring synaptic responses between two neighboring neurons cultured for 3-5 weeks using microelectrodes filled with 1 M potassium acetate (60-80 MΩ) (Mochida et al. 1994a,b; Figure 1C). Neuron pairs were selected by the proximity of their cell bodies. Postsynaptic responses were recorded from one of the neurons when action potentials were evoked in the other neuron by intracellular current pulses passed through the recording electrode. Experiments were carried out at 32°-34°C. Neurons were superfused with modified Krebs' solution consisting of (in mM) 136 NaCl, 5.9 KCl, 5.1 CaCl2, 1.2 MgCl2, 11 glucose, and 3 HEPES (pH 7.4). Data were collected and analyzed on a computer using software written by Dr. L. Tauc (CNRS, France).

C

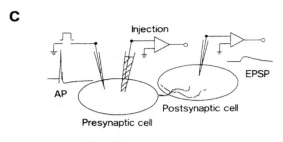

D

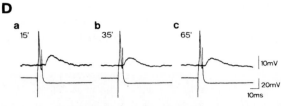

Fig. 1C-D. Recordings of synaptic responses at the synapse formed between superior cervical ganglion neurons (SCGNs) in culture. **C:** Synapse formation between SCGNs was examined by recording synaptic responses with an intracellular microelectrode in response to stimulating a neighboring neuron with a second microelectrode. Injection of agents was achieved by placing a patch-pipette against the cell membrane and applying suction at the designated time. Breakthrough of the membrane patch allows reagents contained within the pipette to diffuse into the presynaptic neuron. To confirm that the agents entered to cells, vital dye, Fast Green FCF, was applied along with the agents. **D:** Postsynaptic responses (EPSPs, *upper traces*) in response to stimulating repetitively (0.05 Hz) a presynaptic neuron by generating action potentials (*lower traces*) at 15 min (**a**), 35 min (**b**) and 65 min (**c**) after the start of stimulation. A gradual decrease in EPSPs for 15-20 min after starting repetitive stimulation was usually observed. However, no significant decline in EPSPs was seen during the subsequent 1-2 h of recording, indicating that the synaptic transmission is stable enough to study the function of proteins expressed in presynaptic terminals in culture

2.4 Synaptic Transmission

The postsynaptic potentials were recorded in the neurons after 1 week in culture (Mochida et al. 1994a). The synaptic potentials increased with time in culture; at 11-14 days in culture, some neurons produced excitatory postsynaptic potentials (EPSPs) that were sufficiently large to generate action potentials. Unidirectional synaptic responses as well as recurrent and reciprocal synaptic responses were recorded in neurons after >2 weeks in culture. The incidence of detecting synaptic potentials were also increased with time in culture; after 2 weeks, EPSPs were detected in 50-80% of the pairs (Mochida et al. 1994a). In some synapses of neuron pairs cultured for 3-5 weeks, relatively stable subthreshold synaptic potentials were recorded in response to repetitive stimuli (0.08-0.2 Hz) (Figure 1D). A gradual decrease in EPSPs for 15-20 min after starting repetitive stimulation was usually observed. However, no significant decline in EPSPs was seen during the subsequent 1-2 h of record, indicating that the synaptic transmission is stable enough to study the function of proteins expressed in presynaptic terminals in culture. In a very few cases, neurons were coupled by electrical synapses. Resting potentials and action potentials were similar to those of intact SCG neurons; -50 to -60 mV for resting potentials and 80 to 100 mV for action potentials, respectively. The input resistances (steady state) of these neurons were around 100 MΩ.

2.5 Microinjection Procedures

For functional studies, intracellular recording techniques were used to measure postsynaptic electrical responses elicited by current injection into a neighboring presynaptic neuron as described whereas whole-cell patch-clamp recording techniques were used for simultaneous injection of reagents into the presynaptic neuron (Mochida et al. 1994a,b, 1995, 1996; Figure 1C). Synaptic transmission was monitored between closely spaced (<5 μm) pairs of neurons for 20-30 min, and then substances were injected into the presynaptic cell body by diffusion from a patch pipette (17-20 MΩ tip resistance). To confirm disruption of the membrane, the membrane potential was also recorded with the injection pipette; 5% Fast Green FCF (Sigma) was applied along with the agents to visualize their entry into the presynaptic cell body through the disrupted membrane and to estimate their amounts in the cell body. The injection pipette was removed 2-5 min after applying suction. Reagents to be injected were dissolved in a solution consisting of (in mM) 150 potassium acetate, 5 Mg^{2+}-ATP, and 10 HEPES (pH 7.4). Disruption of the membrane and diffusion of carrier solution showed no significant effect on transmitter release.

3 Interaction of Synaptic Core Proteins And Calcium Channels At The SCGN Synaptic Terminals

Neurotransmitter release is initiated by Ca^{2+} influx through voltage-gated Ca^{2+} channels (Smith and Augustine, 1988; Robitaille et al. 1990) within 200 μs of the arrival of the action potential at a nerve terminal (Llinás et al. 1981). Thus, it is very likely that the fusion machinery is situated very close to Ca^{2+} channels (Yoshikami et al. 1989). The synaptic plasma membrane proteins syntaxin (Bennett et al. 1992; Inoue et al. 1992; Yoshida et al. 1992) and synaptosome-associated protein of 25 kD (SNAP-25) (Oyler et al. 1989) bind to the synaptic vesicle protein VAMP/synaptobrevin (Trimble 1988) to form a stable synaptic core complex (Söllner et al. 1993; Calakos et al. 1994; O'Conner et al. 1993; Hayashi et al. 1994; Chapman et al. 1994). This complex binds to N-type Ca^{2+} channels (Bennett et al. 1992; Yoshida et al. 1992; Lévêque et al. 1994), and physiological experiments suggest a close association of N-type Ca^{2+} channels with sites of transmitter secretion (Stanley 1993). N-type Ca^{2+} channels bind to the synaptic core complex through a site in the intracellular loop connecting transmembrane domain II and III (Sheng et al. 1994; Rettig et al. 1996) of their α1 subunits (Dubel et al. 1992) in a Ca^{2+}-dependent manner (Sheng et al. 1996).

3.1 N-Type Ca²⁺ Channels Mediate Transmitter Secretion

N-type Ca^{2+} channels are localized in nerve terminals (Robitaille et al. 1990; Westenbroek et al. 1992) and participate in neurotransmitter release in central and peripheral synapses (Tsien et al. 1988; Wu and Saggau 1994; Mintz et al. 1995). At SCGNs synapses, omega-conotoxin GVIA irrversibly blocked synaptic transmitter release from the presynaptic terminals in a dose-dependent manner (Mochida et al. 1995). Synaptic transmitter release was almost completely blocked by GVIA at 5.5 µM in the bathing solution.

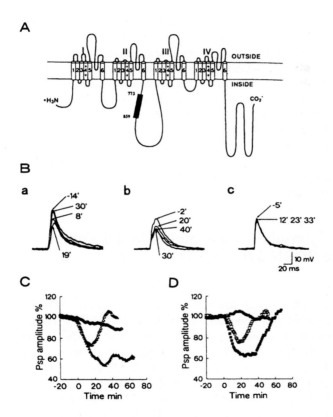

Figure 2. A-D. Synaptic protein interaction site of α1 subunit of N-type Ca^{2+} channels and effects of the fusion proteins on synaptic transmission. **A** Binding of the cytoplasmic domain of α1B to syntaxin 1A. Predicted topological structure of the α1 subunit of class B N-type Ca^{2+} channels. N-type Ca^{2+} channels bind to syntaxin 1A through a site in the intracellular loop connecting domains II and III. (From Sheng et al. 1994). **B** Effects of the fusion protein peptides on synaptic transmission. (**a**) Introduction of 65 µM LII-III (718-963) of α1B. (**b**) Introduction of 130 µM LII-III (718-963) α1B. (**c**) As a control, responses to introduction of L-type Ca^{2+} channel fusion protein peptide, LII-III (670-800) α1S. **C** Normalized average postsynaptic potentials are plotted from 7 experiments with 65 µM LII-III (718-963) α1B (*open triangles*), 7 experiments with 130 µM LII-III (718-963) α1B (*closed triangles*), and 5 experiments with 140 µM LII-III (670-800) α1S (*open squares*). (D) Normalized average postsynaptic potentials following injection of LII-III (718-859) α1B (240 µM, *n=7, open squares*), LII-III (832-963) α1B (200 µM, *n=7, closed squares*), and control carrier solution (*n=7, open triangles*).

3.2 Inhibition of Transmitter Secretion by N-Type Ca^{2+} Channel Fusion Proteins

Syntaxin-binding sites on N-type Ca^{2+} channels were identified by Sheng et al. (1994) and Rettig et al. (1996). Syntaxin specifically interacts with the fusion protein containing the cytoplasmic loop between homologous repeats II and III (LII-III) of the α1 subunit of the class B N-type Ca^{2+} channels (Figure 2). This interaction is mediated by two nearby regions of LII-III of the α1B subunit of N-type Ca^{2+} channels, which are both contained within the fusion protein LII-III (718-963) and are involved in interaction with syntaxin and SNAP-25. The recombinant fusion protein efficiently blocks binding of native N-type Ca^{2+} channels to syntaxin, indicating that this interaction can be disrupted by excess fusion protein containing the synaptic protein interaction ('synprint') site of α1B. Interaction takes place with the C-terminal one-third of syntaxin (residues 181-288), which is thought to be anchored in the presynaptic plasma membrane. This interaction is dependent on Ca^{2+} (Sheng et al. 1996), with maximal binding at 20 μM free Ca^{2+}, near the threshold for transmitter release. Ca^{2+}-dependent interaction of Ca^{2+} channels with the synaptic core complex may be important for Ca^{2+}-dependent docking and fusion of synaptic vesicles.

Results of microinjection experiments revealed functional roles for the N-type Ca^{2+} channel interaction with the synaptic core complex in neurotransmitter secretion (Mochida et al. 1996). Introduction of the fusion protein LII-III (718-963) from α1B reversibly inhibited synaptic transmission (Figure 2). The inhibitory effect was dependent on the injected concentration of LII-III (718-963). With an injection pipette containing 65 μM LII-III (718-963), the maximum decrease in amplitude of EPSPs, -24 ± 4.2% ($n=7$), was observed 10-15 min after starting the injection, and the EPSPs recovered to the control amplitude by 30-40 min after injection. Nearly two-fold greater inhibition in amplitude and duration was observed when a two-fold higher pipette concentration (130 μM) was used. LII-III from the α1S of an L-type Ca^{2+} channel does not interact with syntaxin or SNAP-25 (Sheng et al. 1994). Injection of 140 μM of the fusion protein LII-III(670-800) from α1S produced no significant decrease in EPSPs amplitude, indicating that the inhibitory effect of LII-III (718-963) from the α1B subunit of N-type Ca^{2+} channels is specific. A shorter fusion protein LII-III (718-859), containing only the first region of the two binding sites to synaptic core complex, or a fusion protein LII-III (832-963), containing the second region of the two binding sites, also decreased amplitude of EPSPs. The relative efficiency for inhibition of transmitter release by these fusion proteins, LII-III (718-963) > LII-III (832-963) > LII-III (718-859), was consistent with their rank order of affinity in binding recombinant syntaxin in vitro (Rettig et al. 1996). The decrease in EPSP amplitude caused by introduction of LII-III (718-963) was not dependent on the rate of presynaptic activity. Thus, it is likely that the fusion protein peptides act on the transmitter release process itself rather than by inhibition of vesicle recycling. The simplest interpretation of these results is that inhibition of transmitter release

2.2 Analysis of Fluctuations

The amplitude distribution of EPSCs approximately follows Poisson statistics (Martin and Pilar 1964a; Yawo and Chuhma 1994). Because of the infrequent occurrence of miniature EPSCs, the mean number of quanta in a single EPSC (quantal content, m) was evaluated from the coefficient of variation (Martin 1966; Kuno and Weakley 1972), and the mean size of a single quantum (quantal size, q) was estimated as mean EPSC divided by m.

2.3 Reagents

Pharmacological agents were usually bath-applied through a perfusing line with a constant flow rate. A solution in the chamber (~1 ml) was completely replaced in less than 2 min. Agents used in this study and their sources were as follows: acetylcholine (ACh, Nacalai Tesque, Kyoto, Japan); L-norepinephrine (NE, Nacalai Tesque); epinephrine (Epn, Nacalai Tesque); dopamine (Nacalai Tesque); D,L-normetanephrine (Nacalai Tesque); phentolamine (Research Biochemicals International, Natick, MA, USA); D,L-propranolol (Sigma); yohimbine (Research Biochemicals International); prazocin (Research Biochemicals International); phenylephrine (Research Biochemicals International); clonidine (Research Biochemicals International); (-)isoproterenol (Sigma); 3-isobutyl-1-methylxanthine (IBMX, Sigma); zaprinast (Sigma); Ro-20-1724 (Biomol, Plymouth Meeting, PA, USA); 8-bromoguanosine 3':5'-cyclic monophosphate (8Br-cGMP, Sigma); 8-bromoadenosine 3':5'-cyclic monophosphate (8Br-cAMP, Sigma); Rp-isomer of 8-(4-chlorophenylthio) guanosine-3',5'-cyclic monophosphorothioate (Rp-8pCPT-cGMPS, BioLog, Bremen, Germany); KT5823 (Calbiochem); N^{ω}-nitro-L-arginine methyl ester (L-NAME, Sigma); hemoglobin (Sigma); sodium nitroprusside (SNP, Sigma); (+)-(E)-4-methyl-2-[(E)-hydroxyimino]-5-nitro-8-methoxy-3-hexenamide (NOR1, Dojin, Kumamoto, Japan). ACh was dissolved at a concentration of 100 µM and was puff-applied by a 20ms pressure pulse from a pipette (tip diameter, 1-2 µm) placed within 10 µm from the recording cell. NE, Epn, dopamine, and normetanephrine were bath-applied with equimolar ascorbic acid. All the experiments were done under yellow fluorescent light (wave length, >520nm; EL40SY-F, Matsushita, Kadoma, Japan) to minimize the photodynamic oxidation.

The values in text and figures are mean ± SEM (number of experiments). Statistically significant differences between various parameters were determined using the two-tailed t-test unless specifically noted.

3 Results

3.1 Potentiation of Transmitter Release by Norepinephrine

Transmitter release from the giant presynaptic terminal was measured by recording cholinergic EPSCs from the postsynaptic ciliary neuron. When NE (10 μM) was bath-applied, EPSC was potentiated within 5 min (199 ± 29% of control; $n=6$, $P<0.03$). In contrast, NE produced no significant change in the current produced by ACh (102 ± 6%; $n=5$, $P>0.8$). In a solution with low Ca^{2+} and high Mg^{2+} concentrations an analysis of fluctuations shows that NE decreased the number of synaptic failures, i.e. events, in which presynaptic stimulation did not evoke an EPSC, and increased the occurrence of large EPSCs. Calculations based on these data revealed that NE (10 μM) increased m of the EPSC to 367 ± 150% ($n=7$, $P<0.003$) of control without changing q (96 ± 11%; $n=7$, $P>0.6$). Thus, two different technical approaches show that NE enhances the ability of the presynaptic terminal to release acetylcholine but does not affect the ability of the postsynaptic cell to respond to acetylcholine.

3.2 Paired-Pulse Facilitation

When the presynaptic oculomotor nerve was stimulated by twin pulses at short intervals, the second EPSC was on average larger than the first one (paired-pulse facilitation, PPF, Martin and Pilar 1964b). The mechanism of PPF has been attributed to the enhancement of exocytotic fusion probability as a result of residual Ca^{2+} in the presynaptic terminal (Katz and Miledi 1968; Kamiya and Zucker 1994; Zucker 1996). PPF was accompanied by an increase in m with little change in q (Yawo, unpublished observations). In the present experiments, PPF ratio with pulse interval of 40ms was 1.94 ± 0.13 ($n=5$) when $[Ca^{2+}]_o$ and $[Mg^{2+}]_o$ were 1 and 5 mM, respectively. NE (10 μM) significantly decreased the PPF ratio to 1.52 ± 0.16 with a similar time course to that of the potentiation of the first EPSC ($P<0.001$). Because the size of the readily releasable pool of synaptic vesicles is limited, maneuvers that increase the probability of vesicular exocytosis would deplete the releasable vesicles for the second EPSC and thus, decrease PPF ratio (Debanne et al. 1996; Schulz 1997). This is consistent with the previous observation that NE would increase the exocytotic fusion probability by increasing its Ca^{2+} sensitivity (Yawo 1996).

3.3 Pharmacological Profiles of Noradrenergic Responses

Like NE, Epn (10 μM) also potentiated the EPSC (251 ± 42%; $n=6$, $P<0.02$). After potentiation by Epn, addition of NE no longer potentiated the EPSC (101 ±

3.7 Test of the Involvement of Nitric Oxide (NO)

All these results are consistent with the notion that nanb-AM selectively activates guanylyl cyclase. Two families of guanylyl cyclases have been reported, the soluble guanylyl cyclases which are activated by NO and the particulate guanylyl cyclases which are anchored to the membrane by a single transmembrane domain (Garbers 1992). Is NO involved in the NE-depeendent potentiation of transmitter release? The NE-dependent potentiation was not affected by either the NO synthase inhibitor N^ω-nitro-L-arginine methyl ester (L-NAME, 100 μM, preincubated for 1-2 h) or NO scavenger hemoglobin (30 μM). Moreover, of the NO donors, neither sodium nitroprusside (SNP, 100 μM) nor (±)-(E)-4-methyl-2-[(E)-hydroxyimino]-5-nitro-8-methoxy-3-hexenamide (NOR1, 20 μM) potentiates transmitter release even in the presence of IBMX (100 μM). The experiments using posthatched chicks showed that the LTP of ciliary ganglion synapse was inhibited by L-NAME (100 μM) and induced by SNP (100 μM) (Lin and Bennett 1994). Therefore, the NO-sensitive soluble guanylyl cyclase appears not to be involved in the NE-dependent potentiation of transmitter release in embryonic chick ciliary ganglion.

4 Discussion

4.1 Novel Adrenergic Receptor at the Presynaptic Terminal

The present results indicate that all the following reactions occur in the giant presynaptic terminal of embryonic chick ciliary ganglion: (1) NE activates guanylyl cyclases through a nanb-AM, (2) accumulation of cGMP activates PKG, and (3) PKG phosphorylates a target protein which may be involved in the exocytosis of synaptic vesicles. In addition, the receptor involved in nanb-AM is unique from any known catecholamine receptors with respect to signal transduction mechanisms as well as pharmacological properties. Therefore, it is reasonable to propose the presence of a novel subclass of adrenergic receptors in the giant presynaptic terminal of chick ciliary ganglion. We have called this receptor the 'GC-adrenergic receptor' because it closely couples with guanylyl cyclase activity.

4.2 cGMP-Dependent Potentiation of Transmitter Release

The potentiation of transmitter release from presynaptic terminal is one of the mechanisms of long-term potentiation (LTP). In some synapses, the presynaptic inductin of LTP has been shown to be mediated by cAMP produced by adenylyl cyclase, the activity of which is regulated by both Ca^{2+} and catecholamine receptors (Kuba and Kumamoto 1986; Weisskopf et al. 1994; Chavez-Noriega and Stevens 1994), whereas, cGMP is also involved in the induction of LTP in the hippocampus (Zhuo et al. 1994; Arancio et al. 1995). The activity-dependent Ca^{2+} influx into presynaptic terminal or postsynaptic cell may be the inducing signal of NO synthase (Bredt et al. 1990), a Ca^{2+}-calmodulin-dependent enzyme (Garthwaite 1991). Independently from NO, which activates the soluble guanylyl cyclase, the GC-adrenergic receptor increases cGMP without increasing cytosolic Ca^{2+}. Therefore, like cAMP, cGMP may be regulated by both Ca^{2+} and catecholamines. Catecholamines are widely distributed in the body and are principal neuromodulators in the peripheral and central nervous system. The NO-dependent signals are diffuse and nonspecific and increase cytosolic cGMP, whereas catecholamine-dependent signals are more specific and increase submembranous cGMP. Because of the potent activity of PDEs, the increase of cGMP should be restricted to the region near GC-adrenergic receptors (see the effects of PDE inhibitors). Therefore, although the NO- and catecholamine-dependent signals may convergently activate the PKG near the membrane, they would regulate divergent subcellular mechanisms.

4.3 Mechanisms Involved in Reduction of PPF Ratio

In many synapses including the chick ciliary giant synapse, the magnitude of PPF is negatively correlated with the quantal content of the first response (Debanne et al. 1996; Schulz 1997). The most plausible mechanism seems to be a depletion of docked vesicles for the second release (Debanne et al. 1996; Doburunz and Stevens 1997; O'Donovan and Rinzel 1997). In the present results, the PPF ratio was reduced during the potentiation by NE and cGMP. It is likely that NE increased the exocytotic fusion probability on the first presynaptic release, depleting the available vesicles for the second release. The results of PPF experiments are consistent with the notion that NE enhances the Ca^{2+} sensitivity of the exocytotic fusion probability since it does not increase the Ca^{2+} influx (Yawo 1996).

4.4 The Rate-Limiting Sites of Transmitter Release

Neurotransmitter release is composed of a complex of biochemical reactions that are triggered by Ca^{2+} influx (Fig. 1). In the chick ciliary giant presynaptic terminal, the N-type Ca^{2+} channel is the molecule involved in this rate-limiting step and is modulated by presynaptic receptors, including A_1-adenosine autoreceptor (Yawo and Chuhma 1993), μ-opioid receptor (Yawo et al. 1994) and α_2-adrenergic receptor (Yawo 1996); i.e., at least three receptors convergently regulate the N-type Ca^{2+} channel. These responses are thought to depend on the membrane-delimited action of G-proteins, and are all rapid and readily reversible. Similar mechanisms work for other Ca^{2+} channel subtypes at various synapses

Kuba K, Kato E, Kumamoto E, et al (1981) Sustained potentiation of transmitter release by adrenaline and dibutylyl cyclic AMP in sympathetic ganglia. Nature 291:654-656

Kuba K, Kumamoto E (1986) Long-term potentiation of transmitter release induced by adrenaline in bull-frog sympathetic ganglia. J Physiol (Lond) 374:515-530

Kuno M (1995) The synapse: function, plasticity, and neurotrophism, Oxford University Press, New York, pp79-80

Kuno M, Weakly JN (1972) Quantal components of the inhibitory synaptic potential in spinal motoneurones of the cat. J Physiol (Lond) 224:287-303

Landmesser L, Pilar G (1972) The onset and development of transmission in the chick ciliary ganglion. J Physiol Lond 222:691-713

Langer SZ (1981) Presynaptic regulation of the release of catecholoamines. Pharmacol Rev 32:337-362

Lengyel I, Nichol KA, Sim A TR, et al (1996) Characterization of protein kinase and phosphatase systems in chick ciliary gnaglion. Neuroscience 70:577-588

Lin Y-Q, Bennett MR (1994) Nitric oxide modulation of quantal secretion in chick ciliary ganglia. J Physiol (Lond) 481:385-394

Martin AR (1966) Quantal nature of synaptic transmission. Physiol Rev 46:51-66

Martin AR, Pilar G (1964a) Quantal components of the synaptic potential in the ciliary ganglion of the chick. J Physiol (Lond) 175:1-16

Martin AR, Pilar G (1964b) Presynaptic and post-synaptic events during post-tetanic potentiation and facilitation in the avian ciliary ganglion. J Physiol (Lond) 175:17-30

Marwitt R, Pilar G, Weakly JN (1971) Characterization of two ganglion cell populations in avian ciliary ganglia. Brain Res 25:317-334

McGaugh JL (1989) Involvement of hormonal and neuromodulatory systems in the regulation of memory storage. Annu Rev Neurosci 12:255-287

Meriney SD, Ford MJ, Oliva D, et al (1991) Endogenous opioids modulate neuronal survival in the developing avian ciliary ganglion. J Neurosci 11:3705-3717

Nicoll RA, Malenka RC, Kauer JA (1990) Functional comparison of neurotransmitter receptor subtypes in mammalian central nervous system. Physiol Rev 70:513-565

O'Donovan MJ, Rinzel J (1997) Synaptic depression: a dynamic regulation of synaptic communication with varied functional roles. Trends Neurosci 20:431-433

Schulz PE (1997) Long-term potentiation involves increases in the probability of neurotransmitter release. Proc Natl Acad Sci USA 94:5888-5893

Stanley EF (1994) The calyx-type synapse of the chick ciliary ganglion as a model of fast cholinergic transmission. Can J Physiol Pharmacol 70:S73-S77

Starke K (1981) Presynaptic receptors. Annu Rev Pharmacol Toxicol 21:7-30

Starke K, Gothert M, Kilbinger H (1989) Modulation of neurotransmitter release by presynaptic autoreceptors. Physiol Rev 69:864-989

Takahashi T, Forsythe ID, Tsujimoto T, et al (1996) Presynaptic calcium current modulation by a metabotropic glutamate receptor. Science 274:594-597

Umemiya M, Berger AJ (1994) Activation of adenosine A1 and A2 receptors differentially modulates calcium channels and glycinergic synaptic transmission in rat brainstem. Neuron 13:1439-1446

Weisskopf MG, Castillo PE, Zalutsky RA, et al (1994) Mediation of hippocampal mossy fiber long-term potentiation by cyclic AMP. Science 265:1878-1882

Wu L-G, Saggau P (1997) Presynaptic inhibition of elicited neurotransmitter release. Trends Neurosci 20:204-212

Yawo H (1989) Rectification of synaptic and acetylcholine currents in the mouse submandibular ganglion cells. J Physiol (Lond) 417:307-322

Yawo H (1996) Noradrenaline modulates transmitter release by enhancing the Ca^{2+} sensitivity of exocytosis in the chick ciliary presynaptic terminal. J Physiol Lond 493:385-391

Yawo H, Chuhma N (1993) Preferential inhibition of ω-conotoxin-sensitive presynaptic Ca^{2+} channels by adenosine autoreceptors. Nature 365:256-258

Yawo H, Chuhma N (1994) ω-Conotoxin-sensitive and resistant transmitter release from the chick ciliary presynaptic terminal. J Physiol (Lond) 477:437-448

Yawo H, Chuhma N, Endo K (1994) Modulation of presynaptic calcium channels by neurotransmitters. Biomed Res 15 (suppl)1:9-16

Zhuo M, Hu Y, Schultz C, et al (1994) Role of guanylyl cyclase and cGMP-dependent protein kinase in long-term potentiation. Nature 368:635-639

Zucker RS (1996) Exocytosis: a molecular and physiological perspective. Neuron 17:1049-1055

Synaptic Transmission at the *Drosophila* Neuromuscular Junction: Effects of Metabotropic Glutamate Receptor Activation

D. ZHANG, H. KUROMI, and Y. KIDOKORO

Institute for Behavioral Sciences, Gunma University School of Medicine
Maebashi, 371-8511, Japan

Key words. *Drosophila*, Synaptic transmission, Metabotropic glutamate receptor, Synaptic plasticity, Mutant

Summary: Synaptic transmission at embryonic and larval *Drosophila* neuromuscular junctions has recently been studied in wild-type strains. At the same time isolation of mutants which have defects in various aspects of synaptic transmission is currently actively carried out in many laboratories. These mutants will enable us to dissect underlying molecular events during synaptic transmission. For the effective use of these mutants it is first necessary to examine synaptic transmission in the wild-type embryos and larvae in detail. In this article we describe the effect of metabotropic glutamate receptor activation on neuromuscular synaptic transmission in a wild-type strain. Metabotropic glutamate receptors are found abundantly in rat hippocampus and cerebellum suggesting their important roles in the brain function (Masu et al., 1991). Recently a type of *Drosophila* metabotropic glutamate receptor was cloned and expressed in a cell line, and basic pharmacological properties were described. We examined the effect of glutamate or its agonists on neuromuscular transmission in newly hatched *Drosophila* larvae and found that metabotropic glutamate receptor is operating at this synapse to enhance transmission. The frequency of miniature synaptic currents transiently increased upon puff-application of low concentrations of glutamate in the absence of external Ca^{2+}. Effective glutamate concentrations are lower than those to activate postsynaptic glutamate receptor channels. The similar effects were evoked by a metabotropic glutamate receptor agonists, (S)-4C3HPG or DCG-IV, but not by ionotropic glutamate receptor agonists, NMDA, AMPA or kainate. This effect of glutamate was blocked by a metabotropic glutamate receptor antagonist, MCCG-I. In the presence of external Ca^{2+} the amplitude of nerve-evoked synaptic currents was slightly enhanced by the metabotropic agonist, (S)-4C3HPG, and slightly reduced by the antagonist,

MCCG-I. Mutants which have defects in the metabotropic glutamate receptor activation pathway should manifest partially altered synaptic transmission.

1 Synaptic Transmission at the Embryonic and Larval *Drosophila* Neuromuscular Junction

For elucidation of the molecular mechanism of glutamatergic synaptic transmission the *Drosophila* neuromuscular junction offers an unique opportunity. Many mutant strains that have defects in various aspects of synaptic function are available for detailed analyses (Littleton and Bellen 1995). Some of these mutants are embryonic lethal and do not survive to later stage larvae, but since neuromuscular junctions form before hatching and it is possible to study them at the earliest stage of synapse formation with the patch-clamp technique, this preparation is an excellent model. Basic synaptic properties of embryos and early larvae have been amply described using the patch clamp technique (Broadie and Bate 1993; Kidokoro and Nishikawa 1994; Nishikawa and Kidokoro 1995). Glutamate receptors appear on the myotube surface as clusters at the time of initial neuromuscular junction formation. Interestingly, those clusters are often associated with nuclei. These clusters subsequently disperse and receptors accumulate at the junction (Saitoe et al. 1997).

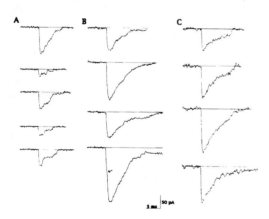

Fig. 1A-C. Spontaneous synaptic currents at stage A (**A**), at stage C (**B**) and at stage E (**C**). Records in **A** were obtained at -60 mV holding potential and others were recorded at -30 mV. An *arrow* in the bottom trace in **B** points at a notch on the rising phase which indicates that this synaptic current was a doublet. In this case two miniature synaptic currents occurred in succession with an interval of about 0.3 ms. Stage A, approximately 16 h after fertilization; stage C, 1-5 hours after hatching; stage E, 1-2 days after hatching. (From Kidokoro and Nishikawa 1994, with permission).

This process of receptor accumulation is similar to that found in *Xenopus* nerve-muscle cultures (Kuromi and Kidokoro 1983; Kuromi et al. 1984). Reflecting this process, the amplitude of miniature synaptic currents increases (Kidokoro and Nishikawa 1994; Nishikawa and Kidokoro 1995) (Figure 1), and the sensitivity to iontophoretically applied glutamate also increases during this period (Broadie and Bates 1993). By the time of hatching functional differentiation of the neuromuscular junction is nearly complete. Morphologically, however, the neuromuscular junction at this stage is still simple compared with that in third-instar larvae. (Atwood et al. 1993; Yoshihara et al. 1997).

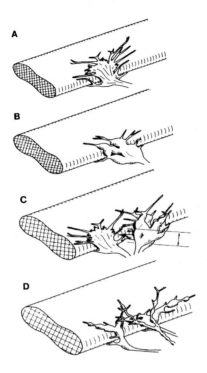

Fig. 2A-D. Diagrammatic representation of steps in development of the junctional aggregate on a single muscle fiber. **A** Growth cone stage. During the growth cone stage, the developing terminal is thin and flat, with long filopodia. The growth cone from a single axon is illustrated. However, it often appears that toward the end of the growth cone stage two or more growth cones are overlapping in the vicinity of the nerve entry point. **B** Prevaricosity stage. The prevaricosities formed by a single axon are shown for simplicity. The growth cone condenses into several recognizable branches that have distinct thickness and rounded contours. Filopodia are shorter. Simple presynaptic specializations form along broad contact regions with the muscle fiber. **C** Junctional aggregate at the prevaricosity stage, about 16.5 to 18.5 h after fertilization. During the prevaricosity stage, several terminals of differing degrees of development are usually found overlapping at the nerve entry point. In this example, a growth cone is shown slightly diverging to the *left*, and two overlapping terminals with prevaricosities are to the *right*. Many additional configurations have been observed. When terminals have entered at the same point, they often remain spatially close, with membrane-to-membrane contact for some distance away from the nerve entry point. This leads to a very complicated appearance when seen at either the light microscope or scanning electron microscope level. **D** At 18-19 h after fertilization, distinct varicosities, swellings with constrictions on either side, resolve from the enlarge branches of the prevaricosity. A single swelling may divide into two or three discrete varicosities. Filopodia are shorter, the elements of the subsynaptic reticulum begin to separate the broad nerve-muscle contact regions, and individual bouton types can begin to be recognized. Subsequent development between this stage and first instar is a matter of degree of development of individual varicosities. (From Yoshihara, et al. 1997, with permission)

Approximately 13 h after fertilization, growth cones contact target muscles. This initial contact occurs almost simultaneously among muscles located in the ventral and dorsal parts of an abdominal segment, although the distances between the ventral nerve cord and muscles are different from muscle to muscle. At 2 to 3 h after this encounter, the growth cone transforms into a transient structure, called 'prevaricosity,' before forming a mature varicosity. Multiple innervation of each muscle fiber occurs at this time. Since this is the period when recognition of the correct target and further differentiation of synapse take place, the prevaricosity stage may include key steps during synapse formation (Yoshihara et al. 1997) (Figure 2). It is thus important to further describe details of synaptic transmission at these early stages.

2 Metabotropic Glutamate Receptors

There are eight types of metabotropic glutamate receptors (mGluRs, 1 through 8), and pharmacologically they are classified into three groups (group I: mGluR1, mGluR5; group II: mGluR2, mGluR3; group III: mGluR4, mGluR7, mGluR8, mGluR6) (Pin and Duvoisin, 1995). Application of mGluR agonists, L-2-amino-4-phosphonobutylate (AP-4), a specific activator of Group III or 1S,3R-1-amino-1,3-cyclopentane-dicarboxylate (1S,3R-ACPD), an activator of Group I and Group II, results in depression of synaptic transmission in a variety of preparations (Pin and Duvoisin 1995). Direct effects through the GTP-binding protein on voltage-gated Ca^{2+} channels are also demonstrated in the slice of rat superior olivary complex (group III, Takahashi et al. 1996). On the other hand, presynaptic potentiating effects of mGluRs have also been demonstrated in various preparations. For example, in cultured Purkinje cells application of an mGluR agonist, quisqualate, a potent activator of group I mGluRs, induced an elevation of internal Ca^{2+} even in the absence of external Ca^{2+} (Yuzaki and Mikoshiba, 1992). They postulated that activation of mGluRs leads to production of IP3 through GTP-binding protein. Then, if these mGluRs are localized in the presynaptic terminal, mobilization of internal Ca^{2+} may cause enhancement of synaptic transmission. Apart from activation of the phospholipase C cascade, mGluRs also modulate the pathway involving cyclic AMP (cAMP) positively in some cases and negatively in many other cases (Pin and Duvoisin, 1995). Thus the effects of mGluR activation on synaptic transmission are extremely diverse, and mGluRs may play important roles in synaptic plasticity.

Recently a type of Drosophila mGluR (DmGluRA) was cloned and expressed in a cell line, and basic pharmacological properties were described (Parmentier et al. 1996). Based on the sequence homology and pharmacological properties DmGluRA was classified as Group II, activation of which leads to indirect effects on synaptic transmission through the cAMP pathway rather than through the phospholipase C cascade (Pin and Duvoisin 1995). Furthermore, DmGluRA was expressed transiently at stage 10 and later at stage 14 to 17 in the CNS. The second period of expression coincides with formation of the glutamatergic neuromuscular junction (Broadie and Bate 1993). Thus, it has been suggested that DmGluRA plays a role in embryonic synapses.

Suppressive Effects of Serotonin on Autaptic Transmission in Cultured Rat Hippocampal Neuron

M. KOGURE[2], O. TAJIMA[2] and K. YAMAGUCHI[1]

[1]Dept. Physiol., [2]Dept. Neuropsychiat., Kyorin Univ. Sch. Med., 6-20-2 Shinkawa, Mitaka, Tokyo 181-8611, Japan

Key words. Serotonin, 5-HT$_{1A}$ receptor, 8-OH-DPAT, Whole-cell recording, Hippocampus

Serotonin (5-hydroxytryptamine, 5-HT) mediates a wide range of physiological functions by activating multiple receptors (Hoyer et al. 1994). Most 5-HT receptors regulate intracellular signal transduction system through GTP-binding proteins (Sanders-Bush and Canton 1995). These intracellular signal transductions mediated by 5-HT receptors of soma-dendritic region of neurons were mostly studied using brain slice preparations (Aghajanian 1995). Though 5-HT receptors distribute not only at the soma-dendritic region but also at the presynaptic terminal region, physiological properties of the presynaptic 5-HT receptors remain unclear. We attempted to analyze the action of 5-HT on synaptic transmission using autapse of the rat hippocampal neuron in culture, as both pre- and postsynaptic action could be analyzed in a single cell, and found that 5-HT suppressed the amplitude of excitatory postsynaptic current (epsc). This action was suggested to be mediated through 5-HT$_{1A}$ receptor at the presynaptic terminal.

Newborn rat (P2-P4) hippocampal neurons were dissociated using papain (9 units ml^{-1}, Worthington) and plated on glial micro-islands (Yamaguchi et al. 1997). Cells were cultured in DME medium supplemented with sera (15%) for the first 3 days, which was then replaced with a serum-free, B27-supplemented (2%, Gibco) DME medium containing 2-amino-5-phosphonovaleric acid (APV, 50 μM, Tocris). Autaptic electrical activities were recorded using whole-cell patch-clamp methods. The bath solution was modified Krebs solutions containing APV (30 μM) and picrotoxin (30 μM); the nternal solution consisted of mainly K-aspartate containing Ca-EGTA, Mg-ATP, GTP and HEPES (Yamaguchi et al, 1997).

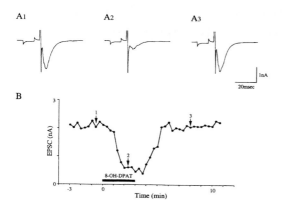

Fig. 1 Effect of 8-OH-DPAT on autaptic epsc. **A, B** Bath-application of 8-OH-DPAT (30 μM) generated a slow suppression of autaptic epsc. A1-A3, current-traces recorded at 1-3 in B, respectively.

Autaptic epscs were recorded from a cell solely grown on a glial micro-island. Bath-application of 5-HT (10 μM) caused suppression of the autaptic epsc amplitude (mean suppression ratio of epsc: 50 ± 11 %, mean ± SEM, n=5). Application of 8-OH-DPAT (3-30 μM), a selective agonist of 5-HT_{1A} receptor, also suppressed the autaptic epsc (Fig. 1). At 10 μM, the mean suppression ratio of epsc was 48 ± 10 % (n=5), which suggested that 5-HT acted mainly through 5-HT_{1A} receptor. Then, the effect of intracellular application of GTP-γ-S, a nonhydrolyzable GTP analogue, was examined. When GTP-γ-S (0.1mM) containing pipette solution was used, 8-OH-DPAT caused almost irreversible suppression of the autaptic epsc. Mediation of G-protein was strongly suggested. Analysis of asynchronous epsc amplitude indicated that the amplitude distribution pattern was not different between those recorded before and during 5-HT action, suggesting that the sensitivity of glutamate receptor is not affected by 5-HT. Thus, decrease in evoked epsc by serotonergic agonists was suggested to be attributable to the presynaptic mechanism.

References

Aghajanian GK (1995) Electrophysiology of serotonin receptor subtypes and signal transduction pathways. In: Psychopharmacology: the fourth generation of progress. Bloom FE, Kupfer DJ (eds) Raven Press, New York, pp 451-460

Hoyer D, Clarke DE, Fozard JR et al (1994) International union of pharmacology classification of receptors for 5-hydroxytryptamine (serotonin). Pharmacol Rev 46:157-203

Sanders-Bush E & Canton H (1995) Serotonin receptors: Signal transduction pathways. In: Psychopharmacology: The fourth generation of progress. Eds. Bloom FE & Kupfer DJ. Raven Press, New York, p.431-441

Yamaguchi K, Takada M, Fujimori K et al (1997) Enhancement of synaptic transmission by HPC-1 antibody in the cultured hippocampal neuron. NeuroReport 8:3641-3644 piperidine dicarboxylic acid, inhibited the second CF-EPSC of the pair proportionately more than the first, suggesting that presynaptic release by the second pulse is decreased. Application of a mGluR agonist, ACPD (Fig. 2A) and a specific mGluR2/3 agonist, DCG-IV, reversively depressed CF-EPSCs by inhibition of presynaptic release. This inhibition was antagonized by a mGluR blocker, MCPG (Fig. 2B). However, MCPG had no effect on PPD, suggesting that presynaptic mGluR2/3 do not contribute to PPD. These results indicate that decreased transmitter release is a major cause of PPD at cerebellar CF to PC synapses (Hashimoto and Kano 1998).

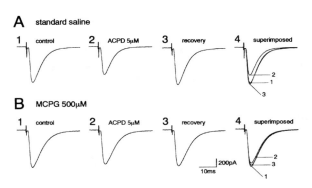

Fig. 2A,B. A Specimen records of CF-EPSCs (average of 5 consecutive responses) taken sequentially (*1*) before, (*2*) during bath application of (1S,3R)-1-aminocyclopentane-1,3-dicarboxylic acid (ACPD, 5 μM), and (*3*) after washing out ACPD (recovery). In *4*, the CF-EPSC traces in *1, 2, 3* are superimposed. **B** Similar to A, but in the presence of (*S*)-α-methyl-4-carboxyphenylglycine (MCPG, 500 μM). Note that MCPG significantly antagonized the ACPD-induced depression of CF-EPSC.

Acknowledgments. This work has been supported by CREST (Core Research for Evolutional Science and Technology) of Japan Science and Technology Corporation (JST) and by grants from the Japanese Ministry of Education, Science, Sports and Culture.

References

Hashimoto K, Kano M (1998) Presynaptic origin of paired-pulse depression at climbing fibre to Purkinje cell synapses in the rat cerebellum. J Physiol (Lond) 506:391-405

Ito M (1984) The cerebellum and neural control. Raven Press, New York

Konnerth A, Llano I, Armstrong CM (1990) Synaptic currents in cerebellar Prukinje cells. Proc Natl Acad Sci USA 87:2662-2665

Adrenaline-Induced Long-Lasting Potentiation of Transmitter Release at Frog Motor Nerve Terminals

S.-M. HUANG, T. AKITA, A. KITAMURA, S. NAKAYAMA and K. KUBA

First Dept. of Physiology, Nagoya University, School of Medicine. Nagoya 466-8550, Japan

Key words. Adrenaline, Transmitter release, Presynaptic terminal, Epp, Mepp, LTP, Intracellular recording, Ca^{2+} measurement

Summary. Adrenaline produced use-dependent, long-term enhancement of the amplitude of end-plate potential (Epp) with a decrease in the coefficient of variation of the amplitude and marked augmentation of tetanic and posttetanic increases in Epp amplitude at frog neuromuscular junctions. It did not, however, affect the frequency of miniature EPP and an impulse-induced rise in intracellular Ca^{2+}. The results suggest that adrenaline increases the readily releasable pool of transmitter in presynaptic terminals.

1 Introduction

Adrenaline has been known to enhance transmitter release at motor nerve terminals, providing the basis for defatiguing effects of sympathetic nerve activity on skeletal muscles (Orbeli 1923; Jenkinson et al. 1968; Kuba 1970; Kuba and Tomita, 1971). The mechanism of the action was suggested to be an increase in the efficiency of exocytosis (Kuba and Tomita 1971). We have further studied the mechanisms of the actions of adrenaline at frog neuromuscular junctions.

2 Methods

End-plate potentials (EPPs) and miniature EPPs (MEPPs) were recorded from frog cutaneus pectoris muscles with intracellular electrodes, while changes in intracellular free Ca^{2+} concentration ($[Ca^{2+}]_i$) in presynaptic nerve terminals were measured by fluorometry of Ca^{2+}-senstive probes, Fura-2, or Oregon Green BAPTA-1 using a photomultiplier or confocal laser scanning microscope. The K salt of Ca^{2+} indicators was loaded from the cut end of the nerve via axonal transport.

3 Results and Discussion

Adrenaline (5-20 μM) increased the amplitude of EPP with a decrease in coefficient of their variation in a low Ca^{2+}, high Mg^{2+} solution. The enhancement of EPP lasted for more than 3 h after the removal of adrenaline. Similar effects of adrenaline were also seen in normal Ringer solution containing d-tubocurarine (3 μM). The amplitude and frequency of MEPPs recorded in normal or high K^+ solutions remained unchanged during and after the application of adrenaline (10 μM). Thus, adrenaline potentiates the impulse-induced release of transmitter. Furthermore, adrenaline (10 μM) did not affect impulse-induced rises in $[Ca^{2+}]_i$ recorded with either a confocal microscope or fluorometry using a photomultiplier, indicating that the facilitatory action of adrenaline does not result from the enhancement of a process to increase $[Ca^{2+}]_i$ by nerve impulses (Huang et al. 1999).

The action of adrenaline was use-dependent. First, it was not seen at a stimulation interval of 12 sec and increased with a decrease in interval up to 2 s (Huang et al. 1999). Second, adrenaline enhanced markedly the initial transient, and moderately enhanced the later sustained phases of potentiation of EPP amplitude produced after a high frequency tetanus (50 Hz for 2 min). EPPs during tetanus were also moderately enhanced by adrenaline (unpublished observations).

There are two possible mechanisms for the facilitatory action of adrenaline: an increase in the amount of transmitter available for release (an increase in the rate of mobilization of transmitter) and enhancement of the efficiency of exocytosis following a rise in $[Ca^{2+}]_i$ by each nerve impulse. If the second mechanism were the case, adrenaline should have enhanced EPPs during tetanus and the action of adrenaline must not have been use dependent. It is most likely that adrenaline enhances impusle-evoked transmitter release via an increase in the readily releasable pool of transmitter. This effect would not occur at rest so that spontaneous release of transmitter could not be enhanced by adrenaline as seen in the absence of actions on MEPP frequency (unless endocytosis is activated).

The findings of the present study differ from the previous observations in three respects: selective enhancement of evoked release of transmitter, long-lasting nature (see Jenkinson et al. 1968; Kuba 1970) and use-dependence. These characteristics of the facilitatory action of adrenaline appear to be appropriate for the defatiguing effects of sympathetic nerve activity on skeletal muscles.

References

Huang SM, Akita T, Kitamura A, Tokuno H, et al (1999) Long-term, use-dependent enhancement of impulse-induced exocytosis at frog motor nerve terminals. Neurosci Res (In press)

Jenkinson DH, Stmenovic H, Whitaker BDL (1968) The effect of noradrenaline on the end-plate potential in twitch fibres of the frog. J Physiol 195:743-754

Orbeli LA (1923) Die sympathetische Innervation der skelettmuskeln. Bull Inst Sci Leshaft 6:194-197

Kuba K (1970) Effects of catecholamines on the neuromuscular junction in the rat diaphragm. J Physiol 211:551-570

Kuba K, Tomita T (1971) Noradrenaline action on nerve terminal in the rat diaphragm. J Physiol 217:19-31

Introductory Review: Synaptic Plasticity and Modulation

T. YOSHIOKA[1] AND K. KUBA[2]

[1]Department of Molecular Neurobiology, School of Human Sciences, Waseda University, Tokyo,
[2]Department of Physiology, School of Medicine, Nagoya University, 65 Tsurumai-cho, Showa-ku, Nagoya 466-8550, Japan

Key words. Synaptic plasticity, Modulation, Long-term potentiation, LTP, Long-term depression, LTD, Hippocampus, Cerebellum, Ca^{2+}, Phosphorylation, Dephosphorylation, mRNA, DNA, Gene knock-out, Memory, Learning

1 Introduction

A change in efficacy of synaptic transmission between neurones occurs in response to their own or other synaptic activity or to the actions of modulators and remains even after the subsidence of its causes (Bliss and Collingridge, 1993). This change, called synaptic plasticity, is thought to be a basis for learning and memory, and could also be one of the important characteristics of synapses in general not necessarily related to the former context. In a broad view, synaptic plasticity is a change in the set point of synaptic strength linking two neurones or the formation or removal of neuronal connections, and so includes developmental processes of neuronal circuitry. Short-term plasticity that lasts for less than tens of minutes is reviewed by Mochida and Kuba in the section on exocytosis and modulation. Long-term synaptic plasticity, represented by long-term potentiation (LTP: Bliss and Lomo, 1973) and long-term depression (LTD: Ito et al. 1982) lasting for more than hours are reviewed here.

2 An overview on the mechanism of LTP and LTD

2.1 Characteristics of LTP and LTD

LTP and LTD usually occur only in the synapses activated by their own pre- or postsynaptic activity (see reviews by Bliss and Collingridge, 1993; Huang et al. 1996). In many instances, the induction of LTP or LTD needs the temporarily close coactivation of several presynaptic inputs, pre- and postsynaptic neurons (Hebbian synapses), pre- or postsynaptic and modulatory inputs, or different transmitter receptors on pre- and/or postsynaptic neurons. For instances, coactivation of NMDA- and AMPA-type glutamate receptors is necessary for the induction of LTP in Schaffer collateral-CA1 pyramidal neurone synapses. The former receptor depolarizes the postsynaptic membrane to relieve voltage-dependent blocking action of Mg^{2+} on the latter, the activation of which leads to the induction of LTP (see Malenka and Nicoll, 1993). Coactivation of AMPA-type and metabotropic glutamate receptors is required for the generation of LTD in parallel fiber-Purkinje neurone synapses in the cerebellum (see Linden, 1994). The mechanisms of LTP and LTD involve a cascade of many processes, whose roles are conventionally categorized into those in induction, expression, and maintenance. Each process could occur at either the pre- or postsynaptic neuron, or at both, and each process could play single or multiple roles (Kuba and Kumamoto, 1990). The time course of LTP (and presumably LTD) is, in general, separated into early and late phases (see Fujita et al. in this section). The early phase seems to be maintained by protein phosphorylation, while the late phase, which is produced by more intense stimulation, is ascribed to protein synthesis. Protein synthesis occurs either by the activation of mRNA in a localized region of synapses or by the activation of DNA in the nucleus and subsequent protein synthesis in the cell body or a region of activated synapses (Frey et al. 1993; Huang et al. 1996).

2.2 Mechanisms of induction and maintenance

A cascade mechanism of LTP or LTD may be summarized into (1) the physiological stimuli given to, or induced in, synapses, (2) the production of second messengers in pre- or postsynaptic neurons, (3) the activation of protein kinases or phosphatases that lead to protein phosphorylation or dephosphorylation, (4) the activation of mRNA or DNA for protein synthesis especially in the late phase of LTP or LTD, and finally (5) changes in the number or state of molecules or structures involved in synaptic transmission. Molecules involved in this cascade have been identified by blocking the function of putative molecules by inhibitors, antibodies, competitive peptides to them, or their geneknockout (Madison et al. 1991; Huang et al. 1996).

3.1.2.2 Evidence for Postsynaptic Expression

There are several findings that could rule out the presynaptic locus of the expression of LTP in CA1 area, leaving the possibility of postsynaptic expression. The rate of the use-dependent blockade of NMDA-receptors by a blocker, MK-801, during the generation of EPSCs may reflect the amount of transmitter release, assuming that the property of the receptor channel remained unchanged. The decay phase of the NMDA component of EPSC should be shortened, if the use-dependent blockade of NMDA receptors is enhanced by an increase in the release of glutamate, but this was not the case (Manabe and Nicoll, 1994). A similar lack of changes in the rate of a use-dependent blockade of AMPA-receptor channel by a use-dependent antagonist, HPP-SP, during LTP was seen in the CA1 region of mice lacking gluR2 subunits (Mainen et al. 1998). The amount of glutamate released from presynaptic terminals is measured by recording a glutamate transporter current at the cell membrane of astrocytes induced by synaptic stimulation. This transporter current remained unchanged after the induction of LTP, although this current was increased concomitantly with EPSCs by paired pulses or increasing the extracellular Ca^{2+} concentration (Diamond et al. 1998; Lüscher et al. 1998).

There are several lines of evidence for the postsynaptic expression of LTP. First, glutamate-induced currents of postsynaptic neurons were increased after the induction of LTP (Davies et al. 1989). Second, the amplitude of miniature EPSCs was elevated during the generation of LTP of EPSC (Manabe et al. 1992). Third, unitary conductance of AMPA receptor channel measured from the noise analysis of the decay phase of EPSC was increased during the generation of LTP (Benke et al. 1998). This is consistent with the finding of phosphorylation of AMPA-type glutamate receptor by CAMKII during LTP (Barria et al. 1997). Finally, there is evidence for the insertion of AMPA receptor channels into the postsynaptic membrane via membrane fusion processes during LTP. LTP was blocked by intracellular injection of *N*-ethylmaleimide, a blocker of NSF (which is an ATPase involved in membrane fusion), botulinum toxin, a blocker of SNAP-25 (which is a protein involved in membrane fusion), and a peptide that constitutes a part of SNAP-25. On the other hand, injection of SNAP-25 produced enhancement of EPSCs for more than 60 min (Lledo et al. 1998).

3.1.2.3 Silent synapses

During development of hippocampal synapses, NMDA receptors appear at the postsynaptic membrane as early as at the beginning of postnatal days and are functional through the later period, while most synapses lack AMPA receptors at early postnatal days (Petralia et al. 1999; Liao et al. 1999). These synapses are not functional, because NADA receptor is not activated unless membrane depolarization by AMPA receptor removes the Mg^{2+}-dependent blockade. In this context, this can be called a "silent synapse" or "deaf synapse". The number of silent synapses decreases as development progresses (Petralia et al. 1999; Liao et al. 1999).

In the synapses having no AMPA receptors at day 16-18, pairing low-frequency presynaptic stimulation (0.3-2.0 Hz, 100 pulses) with postsynaptic depolarizatin to −10 mV resulted in the appearance of EPSCs at −65 mV with no change in EPSCs (NMDA receptor-mediated) at +60 mV (Liao et al. 1995). This induction of AMPA responses was input specific (Liao et al. 1995) and dependent upon the activation of NMDA receptor and a rise in $[Ca^{2+}]_i$, since the induction was blocked by APV and intracellular EGTA injection. AMPA receptors appeared within minutes of pairing after postnatal day 2 and lasted for another 30 min (Durand et al. 1996). In line with this, the density and shape of dendritic spines are maintained by the activation of AMPA receptors by spontaneous release of glutamate (Mckinney et al. 1999).

A recent study suggests the existence of presynaptically silent synapse, "mute synapse", whose presynaptic boutons is not functional, but becomes functional after stimulation that produces the late phase of LTP. The application of an analogue of cyclic AMP, Sp-cAMP, that produces the late phase of LTP, increased the number of boutons stained with FM1-43 in response to nerve stimulation, although there was no change in the number of boutons. This effect was blocked by antagonists of the cyclic AMP pathway, Rp-cAMP, blockers of NMDA and AMPA receptors, APV and CNQX, and an ihibitor of protein synthesis, anisomycin (Ma et al. 1999).

3.1.2.4 Formation of new synapses

The recruitment of new sites of synaptic transmission was suggested to occur during the late phase of LTP based on the detailed analysis of EPSCs in synapses between a pair of presynaptic CA3 and postsynaptic CA1 neurons by Bolshakov et al. (1997). During the late phase of LTP induced by Rp-cAMP, single quantal EPSCs were changed to multi-quantal EPSCs without a change in the quantal size, but with a reduction of the failure rate. The quantal content of EPSCs was increased by raising the external Ca^{2+} during, but not before the induction of, the late phase of LTP. Furthermore, there was no change in the fractions of the NMDA- and AMPA components of EPSC during LTP, ruling out the involvement of silent synapses. A drawback to the conclusion of formation of new synapses is an increase in the amplitude of miniature EPSCs under this condition. This was explained by coincident releases of multiple quanta by an unknown mechanism, since multiquantal MEPSCs disappeared, when they were desynchronized by Sr^{2+} in place of Ca^{2+}.

LTD induced by phosphodiesterase inhibitors was blocked by chelerythrine, a blocker of PKC, which was applied intracellularly in Purkinje neurons (Hartel, 1996b). LTD produced by glutamate and membrane depolarization of a Purkinje neuron was blocked by external application of blockers of PKC, RO-31-8220 and calphostin C, and internal application of a PKC inhibitory peptide [glu^{27}]PKC(19-36). Furthermore, an activator of PKC, phorbol-12,13-diacetate, caused a marked long-lasting depression of AMPA-mediated currents (Crepel and Krupa, 1988; Linden and Connor, 1991). Glial fibrillary acidic protein (GFAP), an intermediate filament protein expressed in astrocytes, appears to play an essential role in the induction of LTD. Induction of LTD was impaired in mutant mice, which has no GFAP (Shibuki et al. 1996).

There is no doubt that a decrease in the sensitivity of the postsynaptic membrane to glutamate is the mechanism of expression of LTD in the cerebellum, as found in the original study (Ito et al. 1982). This idea was supported by a decrease in glutamate- or AMPA-induced currents in Purkinje neurons. The reduction of glutamate sensitivity could be produced by a change in property of AMPA-type glutamate receptor that leads to the reduction of the open probability or conductance, or downregulation of the receptor. In support of the former possibility, Nakazawa et al. (1995) reported the persistent phosphorylation of AMPA-type glutamate receptor subunits following the application of 8-bromo-cyclic GMP, dibutyryl cyclic GMP (but not phorbol 12,13-diacetate) or calyculin A. This suggests the direct role of G-kinase in the mechanism of a change in glutamate sensitivity of the postsynaptic membrane. Induction of LTD was impaired in mice defective in the glutamate receptor channel δ2 subunit (Kashiwabuchi et al. 1995), indicating the involvement of this subunit in the expression of LTD.

If we assume that LTD plays a role in motor leaning, then we think that learning (and memory) may result from changes in the efficiency of synaptic transmission, which will be regulated by the modification of intracellular second messengers. We can also assume that memory storage can result from alterations in existing proteins near the synapse. Recently, Alkon et al (1998) proposed that associative learning, classical conditioning, in Hemissenda correlates with changes in neuronal voltage dependent K^+ current, protein kinase C-mediated phosphorylation and synthesis of small molecular weight G protein, calexcitin (CE). If we compare the mechanism of classical conditioning in Hermissenda with that of LTD in the cerebellum, the final stage for classical conditioning is the reduction of voltage dependent K^+ channel conductance, while that for LTD is a decrease in AMPA current. Activation of PKC can explain both classical conditioning and LTD, because PKC blockers inhibit classical conditioning and LTD. PKC can phosphorylate small G-protein (CE) and the phosphorylated CE reduces A-channel conductance. If the phosphorylated CE can reduce the AMPA channel conductance (not proved yet), studies on molecular mechanisms of learning and memory will develop to a new stage.

References

Aiba A, Kano M, Chen C, et al (1994) Deficient cerebellar long-term depression and impaired motor learning in mGluR1 mutant mice. Cell 79:377-388

Alkon DL, Nelson TJ, Zhao W, et al (1998) Time domains of neural Ca^{2+} signaling and associative memory: steps through a calexcitin, ryanodine receptor, K+ channel cascade. Trends Neurosci 21:529-537

Baranes D, Lederfein D, Huang Y-Y, et al (1998) Tissue plasminogen activator contributes to the late phase of LTP and to synaptic growth in the hippocampal mossy fiber pathway. Neuron 21:813-825

Barria A, Muller D, Derkach V, et al (1997) Regulatory phosphorylation of AMPA-type glutamate receptors by CaM-KII during long-term potentiation. Science 276:2042-2045

Bashir ZI, Bortolotto ZA, Davies CH, et al (1993) Induction of LTP in the hippocampus needs synaptic activation of glutamate metabotropic receptors. Nature 363:347-350

Bekkers JM, Stevens CF (1990) Presynaptic mechanism for long-term potentiation in the hippocampus. Nature 346:724-729

Benke TA, Lüthi A, Isaac JTR, et al (1998) Modulation of AMPA receptor unitary conductance by synaptic activity. Nature 393:793-797

Berridge MJ (1998) Neuronal calcium signaling. Neuron 21:13-26

Bito H, Deisseroth K, Tsien RW (1996) CREB phosphorylation and dephosphorylation: a Ca^{2+}-and stimulus duration-dependent switch for hippocampal gene expression. Cell 87:1203-1214

Bliss TVP, Collingridge GL (1993) A synaptic model of memory: long-term potentiation in the hippocampus. Nature 361:31-39

Bliss TVP, Lomo T (1973) Long-lasting potentiation of synaptic transmission in the dentate area of the anesthetized rabbit following stimulation of the perforant path. J Physiol 232:331-356

Bolshakov VY, Sigelbaum SA (1995) Regulation of hippocampal transmitter release during development and long-term potentiation. Science 269:1730-1734

Bolshakov VY, Golan H, Kandel ER, et al (1997) Recruitment of new sites of synaptic transmission during the cAMP-dependent late phase of LTP at CA3-CA1 synapses in the hippocampus. Neuron 19:635-651

Bortolotto ZA, Bashir ZI, Davies CH, et al (1994) A molecular switch activated by metabotropic glutamate receptors regulates induction of long-term potentiation. Nature 368:740-743

Bredt DS, Snyder SH (1990) Isolation of nitric oxide synthatase, a calmodulin-requiring enzyme. Proc Natl Acad Sci USA 87:682-685

Conquet F, Bashir ZI, Davies CH, et al (1994) Motor deficit and impairment of synaptic plasticity in mice lacking mGluR1. Nature 372:237-243

Crepel F, Jaillard D (1990) Protein kinases, nitric oxide and long-term depression of synapses in the cerebellum. Neuroreport 1:133-136

Crepel F, Krupa M (1988) Activation of protein kinase C induces a long-term depression of glutamate sensitivity of cerebellar Purkinje cells. An *in vitro* study. Brain Res 458:397-401

Davies SN, Lester RAJ, Reyman KG, et al (1989) Nature 338:500-503

Daniel H, Hemart N, Jaillard D, et al (1993) Long-term depression requires nitric oxide and guanosine 3':5' cyclic monophosphate production in rat cerebellar Purkinje cells. Eur J Neurosci 5:1079-1082

298   T. Tsumoto et al.

study, we addressed these questions in two kinds of preparations, visual cortical slices and cultured solitary neurons obtained from the visual cortex of neonatal rats. Part of the data presented here has been published previously (Akaneya et al. 1997).

2 Methods

2.1 Preparation of Slices

Sprague-Dawleys rats, aged from 15 to 25 postnatal days, were deeply anesthetized with ketamine (30 mg/kg, i.p.), and then killed by cervical dislocation. Procedures for preparing and maintaining slices of the visual cortex were detailed previously (Akaneya et al. 1997). The composition of incubation and perfusion medium of the slices was as follows (in mM): NaCl, 124; KCl 5; KH_2PO_4, 1.2; $MgSO_4$, 1.3; $CaCl_2$, 2.4; $NaHCO_3$, 26; and glucose, 10. All the recordings were performed at $31 \pm 1°C$. For stimulation of afferents to layer II/III of the cortex, a bipolar stimulating electrode was placed in layer IV of the cortex (see Fig. 1, top). To record field potentials evoked by test stimulation of layer IV, glass micropipettes filled with 0.5 M sodium acetate containing 2% pontamine sky blue (resistance <10 $M\Omega$) were inserted into layer II/III of the cortex. In part of the experiments, excitatory postsynaptic currents (EPSCs) were recorded from pyramidal cell-like neurons in layer II/III of the cortex through whole-cell patch-clamp electrodes. These electrodes (resistance 4-8 $M\Omega$) were filled with a solution containing (in mM): potassium gluconate, 130; KCl, 10; N-2-hydroxyethylpiperazine-N'-2-ethanesulfonic acid (HEPES), 10; MgATP, 3; Na_2 GTP, 0.5; and adjusted to pH 7.2 by KOH. Recorded neurons were voltage-clamped at -70 mV with a patch-clamp amplifier (Axopatch 200A or 200B, Axon, I, Foster city, USA). Membrane potentials were corrected for the liquid junctional potentials. The fast transient capacitive currents for recording electrodes were canceled, and the series resistance was less than 20 $M\Omega$.

Test pulses of 0.1-ms width were delivered at 0.1 Hz to layer IV, and the stimulus intensity was adjusted to about 1.5 times the threshold to elicit the postsynaptic component of responses in case of field potential recording and 1.2-1.5 times the threshold for EPSCs in case of whole-cell recording. After having confirmed that test shocks induced responses with almost constant amplitude for 5-10 min, human recombinant BDNF (provided from Regeneron, P, Tarry Town, NY, USA), human recombinant NGF (purchased from Boehringer Mannheim, Mannheim, Germany) or K252a (provided from Kyowa Hakko, Tokyo, Japan), was applied to slices for 20 or 30 min through the perfusion medium. Procedures for preparation of these substances were detailed previously (Akaneya et al. 1997). The rate of perfusion of the medium was 200 ml/h throughout all the experiments.

2.2 Culture Preparation

Solitary neurons were cultured by using conventional microisland methods (Kimura et al. 1997). A piece of visual cortex was removed from neonatal rats, enzymatically dissociated with papain (20 U/ml), and triturated with a fire-polished glass pipette. Neurons were plated on previously prepared glial islands grown on collagen dots placed on an agarose sheet, and were grown in a solution based on Eagle's minimum essential medium supplemented with 5% rat serum. A whole-cell patch electrode (resistance, 5-10 MΩ) was used to record synaptic currents in a voltage-clamp mode (see Fig. 2, top left). The neurons were clamped at -70 mV for recording EPSCs unless otherwise noted. A short voltage step from the holding potential to +10 mV for 2 ms was applied to elicit a somatic Na^+ spike. The perfusing solution, unless otherwise noted, contained the following (in mM): NaCl, 150; KCl, 4; $CaCl_2$, 2; $MgSO_4$, 1; HEPES, 10; glucose, 10 (pH 7.4). Glycine (10 μ M) was added to the medium. Osmolality was adjusted to 320-325 mOsm by adding sucrose when necessary. Neurotrophins and drugs such as kynurenic acid, a nonselective antagonist for ionotropic glutamate receptors and 6-cyano-7-nitroquinoxaline-2,3-dione (CNQX), a selective antagonist for non-N-methyl-D-aspartate receptors, were applied with a local perfusion system equipped with a Y-shaped tube. The electrode solutions mostly used were as follows (in mM): Cs-methane sulfonate, 130; HEPES, 10; CsCl, 10; EGTA, 0.5; MgATP, 5; and Na_2GTP, 1.

The amplitude of stimulus-evoked EPSCs was evaluated by subtracting the mean value of the baseline for 5 ms just before the stimulus artifact from the peak value of the EPSCs. For spontaneous EPSCs, data were searched with a manual peak detection procedure that meets the criteria of peaks higher than a threshold of 3 times the standard deviation (SD) of the baseline noise level and with a faster rise than the decay phase. The difference between the current from the baseline (2 ms immediately before the onset of each event) and the peak (average of 3 points) was then measured manually. In some experiments events were detected with the use of an event detection program that is a built-in component of the AxoGraph 3.0 or 3.5. Baseline noise was calculated as a difference between the mean of the equivalent baseline periods and an average of three points 1 ms apart from the end of the baseline period.

CNQX and their reversal potentials were near 0 mV. In the present analysis we did not include neurons in which evoked synaptic currents were not unambiguously identified as EPSCs on the aforementioned basis. All the neurons thus analyzed also showed spontaneous events without somatic activation (spontaneous EPSCs). In most of the neurons tested, BDNF at the concentration of 200 ng/ml enhanced the frequency of spontaneous EPSCs during the whole period of application, although it did not have such a sustained, potentiating effect on evoked EPSCs. An example of this is shown in Fig. 2.

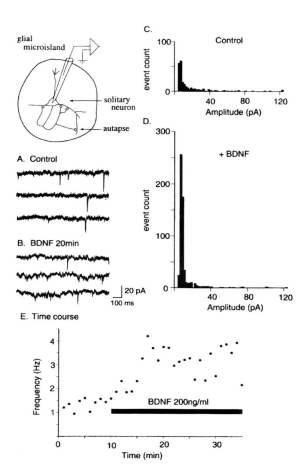

Fig. 2A-D. Increase in frequency of spontaneous EPSCs by BDNF. At top left is shown a schematic depiction of preparations of a solitary neuron. **A, B** Examples of spontaneous events 5 min before and 20 min after the initiation of BDNF application, respectively. Downward deflection represent spontaneous EPSCs. **C, D** Amplitude distribution of spontaneous EPSCs 5-8 min before and 20-23 min after the BDNF application. Total number of events in C and D is 245 and 615, respectively. E Time course of frequency of spontaneous EPSCs. The frequency was measured every 60 s. BDNF at 200 ng/ml was applied to the solitary neuron during the period indicated by the *horizontal bar.*

In the control condition before application of BDNF, this neuron generated spontaneous EPSCs with the mean frequency of 1.4 Hz (Fig. 2A). The distribution of their peak amplitudes was skewed toward larger values so that the median value (5.9 pA) was substantially different from the mean (15.7 pA). A few minutes after starting the application of BDNF, the frequency seemed to increase slightly, although this change was not so remarkable (Fig. 2E). Four to 5 min later, the increase became prominent and this change lasted for the rest of the observation period (Fig. 2B, E). In contrast with such a dramatic change in the frequency, the amplitude of spontaneous EPSCs was not significantly changed (Fig. 2C,D). The median and mean values of the amplitude of spontaneous EPSCs 20-23 min after starting the BDNF application were 6.0 and 9.1 pA, respectively. The former value was almost identical to the control before the BDNF application, although the latter value seemed to be smaller. When the two amplitude distributions were compared statistically, there was no significant difference between them ($P=0.37$, Mann-Whitney U test).

4 Discussion

4.1. Long-Lasting Potentiation of Field Responses by BDNF

In the present study, we found that BDNF at the concentration of 200 ng/ml potentiated layer II/III responses to test stimulation of layer IV, and this potentiation lasted at least for 30 min after cessation of the application. Furthermore, this action of BDNF was blocked by the inhibitor for receptor tyrosine kinases, K252a. These results seem to be generally consistent with the previous reports on hippocampal and neuromuscular synapses, although the effective concentration of BDNF and the duration of the effect are not exactly the same (Lohof et al. 1993; Kang and Schuman 1995; Levine et al. 1995; Figurov et al. 1996; Stoop and Poo 1996). Recently it was reported that BDNF attenuated $GABA_A$ receptor-mediated inhibitory responses, but did not potentiate EPSCs of pyramidal neurons in the CA1 area of rat hippocampus (Tanaka et al. 1977). This suggests that the enhancement of field potentials by BDNF in the present and some of the previous studies might be due to indirect effects following the attenuation of inhibition. This possibility seems unlikely at least in visual cortical slices, however, because the initial slope of EPSCs evoked in layer II/III neurons by test stimulation of layer IV was also enhanced by BDNF. It is unlikely that the initial slope of the EPSCs was contaminated by inhibitory currents in this type of preparations. In fact, Carmignoto et al. (1997) also reported that BDNF induced a rapid and persistent increase in the amplitude of EPSCs of layer III-V neurons in visual cortical slices of young rats. Thus, the difference between the present and the Tanaka et al. (1997) results may be due to differences in structure, use of visual cortex versus hippocampus, or to differences in other experimental conditions. For example, there is a difference in the velocity of BDNF application that is reported to be critical for the presence of the potentiating effect in hippocampal slices (Kang et al.1996).

4.2 Enhancing Action of BDNF on Frequency of Spontaneous EPSCs in Solitary Neurons

In most of the solitary neurons with autapses, we found that BDNF persistently increased the frequency of spontaneous EPSCs, but did not enhance their amplitude. The median value of amplitude of spontaneous EPSCs 20-23 min after the initiation of BDNF application was almost the same as that before the application in the cell shown in Fig. 2. Also, the amplitude distribution after the application was not significantly different from that before the application. Furthermore, the amplitude of evoked EPSCs was not markedly changed by the application of BDNF. These results altogether suggest that the site of the BDNF action on spontaneous EPSCs is presynaptic, since it is unlikely that any change in postsynaptic sites results in no change in amplitude of spontaneous and evoked EPSCs. This interpretation seems consistent with the previous reports regarding the effects of BDNF on synaptic transmission in primary cell cultures from neural tubes and muscles of *Xenopus*, and from hippocampus of fetal or neonatal rats (Lohof et al. 1993; Leßmann et al. 1994; Levine et al. 1995; Stoop and Poo 1996).

If the probability of transmitter release from presynaptic terminals is increased by BDNF, evoked EPSCs also are expected to be enhanced. As mentioned, however, the amplitude of evoked EPSCs was not significantly enhanced in the present study. Such a seemingly different action of BDNF on evoked from spontaneous EPSCs may be accounted for by the following two possibilities. (1) Mechanisms involved in transmitter release in response to terminal depolarization following somatic activation may be different from those in spontaneous release, as suggested previously (Kimura et al. 1997). BDNF might effectively act only on the latter mechanisms. (2) In solitary neurons EPSCs were always elicited in conjunction with postsynaptic depolarization, which must induce an influx of Ca^{2+} into postsynaptic sites. This procedure in turn would lead to long-lasting potentiation of autaptic transmission, because presynaptic activation coupled with an increase in postsynatpic Ca^{2+} is known to induce LTP (see Tsumoto 1992). In most autapses, therefore, the probability of transmitter release in response to somatic activation would already be at or near the saturated level at the time of BDNF application so that the potentiation of amplitude of evoked EPSCs would not further be induced. This possibility also might account for the difference in the effect of BDNF on evoked EPSCs between solitary neuron preparations and slice preparations in the present study. Obviously, further experiments are necessary to test these possibilities.

Ackowledgments. We express many thanks to Regeneron Pharmaceutical Co. and Kyowa Hakko Kogyo Co. for kind gifts of human recombinant BDNF and K252a, respectively.

References

Akaneya Y, Tsumoto T, Kinoshita S, et al (1997) Brain-derived neurotrophic factor enhances long-term potentiation in rat visual cortex. J Neurosci 17:6707-6716

Cabelli RJ, Hohn A, Shatz CJ (1995) Inhibition of ocular dominance column formation by infusion of NT-4/5 or BDNF. Science 267:1662-1666

Carmignoto G, Canella R, Candeo P, et al (1993) Effects of nerve growth factor on neuronal plasticity of the kitten visual cortex. J Physiol (Lond) 464:343-360

Carmignoto G, Pizzorusso T, Tia S, et al (1997) Brain-derived neurotrophic and nerve growth factor potentiate excitatory synaptic transmission in the rat visual cortex. J Physiol (Lond) 498:153-164

Castrén E, Pitkänen M, Sirviö J, et al (1992) The induction of LTP increases BDNF and NGF mRNA but decreases NT-3 mRNA in the dentate gyrus. NeuroReport 4:895-898

Figurov A, Pozzo-Miller LD, Olafsson P, et al (1996) Regulation of synaptic responses to high-frequency stimulation and LTP by neurotrophins in the hippocampus. Nature 381:706-709

Kang H, Jia LZ, Suh KY, et al (1996) Determination of BDNF-induced hippocampal synaptic plasticity: role of the TrkB receptor and the kinetics of neurotrophin delivery. Learn Mem 3:188-196

Kang H, Schuman EM (1995) Long-lasting neurotrophin-induced enhancement of synaptic transmission in the adult hippocampus. Science 267: 1658-1662.

Kim HG, Wang T, Olafsson P, Lu B (1994) Neurotrophin 3 potentiates neuronal activity and inhibits gamma-aminobutyratergic synaptic transmission in cortical neurons. Proc Natl Acad Sci USA 91:12341-12345

Kimura F, Nishigori A, Shirokawa T, et al (1989) Long-term potentiation and N-methyl-D-aspartate receptors in the visual cortex of young rat. J Physiol (Lond) 414:125-144

Kimura F, Otsu Y, Tsumoto T (1997) Presynaptically silent synapses: spontaneously active terminals without stimulus-evoked release demonstrated in cortical autapses. J Neurophysiol 77:2805-2815

Knüsel B, Hefti F (1992) K-252 compounds: modulator of neurotrophin signal transduction. J Neurochem 59:1987-1996

Korte M, Carroll P, Wolf E, et al (1995) Hippocampal long-term potentiation is impaired in mice lacking brain-derived neurotrophic factor. Proc Natl Acad Sci USA 92: 8856-8860

Leßmann V, Kottmann K, Heumann R (1994) BDNF and NT-4/5 enhance glutamatergic synaptic transmission in cultured hippocampal neurones. NeuroReport 6:21-25

Levine ES, Dreyfus CF, Black, IB et al (1995) Brain-derived neurotrophic factor rapidly enhances synaptic transmission in hippocampal neurons via postsynaptic tyrosine kinase receptor. Proc Natl Acad Sci USA 92:8074-8077

Lohof AM, Ip NY, Poo MM (1993) Potentiation of developing neuromuscular synapses by the neurotrophins NT-3 and BDNF. Nature 363:350-353

Maffei L, Berardi N, Domenici L, et al (1992) Nerve growth factor (NGF) prevents the shift in ocular dominance distribution of visual cortical neurons in monocularly deprived rats. J Neurosci 12:4651-4662

McAllister AK, Lo DC, Katz, LC (1995) Neurotrophins regulate dendritic growth in developing visual cortex. Neuron 15:791-803

Patterson SL, Grover LM, Schwartzkroin PA, et al (1992) Neurotrophin expression in rat hippocampal slices: a stimulus paradigm inducing LTP in CA1 evokes increases in BDNF and NT-3 mRNAs. Neuron 9:1081-1088 kinetics of AMPA receptor channels, and (3) the number of functional channels in the postsynaptic membrane may decrease. It is suggested from the present experiment that (1) the number of the receptor channels in the postsynaptic membrane and (2) the affinity of glutamate to the receptor decrease during LTD and that these processes could be caused by modification of receptor channels (Kojima and Ileva 1997; Kojima et al. 1997).

2 Methods

Current recordings were carried our from Purkinje cells, identified by ID-DIC video microscopy, of 200-μm-thick slice preparations of 14- to 21-day-old Wistar rats that were continuously perfused with normal Ringer's solution containing 20 μm picrotoxin for inhibition of miniature inhibitory synaptic currents, using patch-clamp technique in whole-cell and single-channel recording mode (Edwards et al. 1989, Hamill et al. 1981). Monopolar stimulating electrodes were placed in the middle of the molecular layer and the edge of the white matter on the surface of a slice preparation to evoke the synaptic currents by stimulation of parallel and climbing fibers, respectively (Llano et al. 1991). Currents were recorded with Axopatch 1-D patch clamp amplifier and filtered with a 50- to 10-kHz bandwidth (-3dB), with 8-pole low-pass Bessel filter. Data collection and analysis were conducted using pClamp6 system and S.Traynelis' software (Traynelis et al. 1992). Mean waveform of EPSC was scaled up to the peak of each EPSC and then subtracted for the calculation of variances around a mean waveform. Amplitude of EPSC was divided into 100 sections, and mean currents and variances were calculated for each section. The relationship between mean current values and variances was fitted by a parabolic function, and the slope of the curve at the initial point gives a single-channel current through a AMPAR channel which mediates the synaptic current in the synapse between a parallel fiber to Purkinje cell. The recorded EPSCs were filtered at a frequency of 10 kHz and then sampled at 40 kHz for the nonstationary noise analysis (nSNA). The possible filtering effect of the dendrites of cells were estimated qualitatively by calculating a weak correlation between rise time constants and amplitudes of EPSCs. Single-channel current recording was carried out from the somata of Purkinje cells by a cell-attached patch clamp configuration. The patch pipettes filled with normal Ringer's solution had been added with 20 μm picrotoxin and 1 mM of AMPA-HBr.

Single-channel current data recorded were replayed from magnetic tape, amplified, and filtered at a frequency of 1-4 kHz, continuously sampled at 10-40 kHz onto a 486-based PC computer. The duration of open time and closed period in the data record was measured using the method of 50% threshold fitting after first estimating the amplitude of channel opening using manually controlled cursors placed on the data display. Histograms of the distribution of open times and shut times were constructed for display and evaluation of the data. In most cases the distribution of log (duration) is used for display purposes, with a square root transformation of the ordinate. Distributions were fitted with sum of several exponential components where appropriate (Colquhoun and Sigwarth 1995; Sigwarth and Sine 1987).

3 Results

Synaptic currents recorded at somata of the Purkinje cells by stimulating the parallel fibres at 0.5 Hz were fitted by two exponential functions, a single exponential for rising phase with time constants 2.21 ± 0.18 ms (SD, n=10) and another single exponential for decay phase with time constants 10.68 ± 0.08 ms (SD, n=10), respectively.

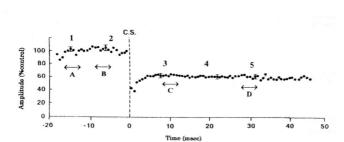

LTD of EPSCs evoked by parallel fibre stimulation

Fig. 1. Long-term depression of EPSCs evoked by parallel fiber stimulation. Long-term depression of EPSCs evoked by PF activation induced by conjunctive stimulation of parallel and climbing fibres at a frequency of 1 Hz for 5 mins (300 pulses). The amplitudes of EPSCs decreased to about 60% of those of control EPSCs recorded for 10 min before the conjunctive treatment. Each trace was an average of 5 consecutive traces. Cell input resistance and series resistance of the recording pipette were monitored every 2 min during the time which the recording was carried out and remained almost unchanged Cell was voltage clamped at a holding potential of –80 mV.

The waveforms of the EPSCs could still be fitted by two exponentials independent of the membrane potential. Although several factors are involved in determining the decay phase of synaptic currents, the deactivation and desensitization mainly contribute to the decay phase of EPSCs and shaped a single exponential function of decay (Barbour et al. 1994; Jonas and Spruston 1994). By conjunctive stimulation (C.S.) of parallel and climbing fibers for 5 min at a frequency of 1 Hz (300 pulses), we observed LTD of synaptic transmission between PF and PC shown in Figure 1. The amplitudes of EPSCs decreased to about 65% of those of control values obtained before C.S. (65.8 ±10.8% [n=10]), and continued at this low level for a period of 30-150 min until the end of the recording. Each EPSC recorded before and after the C.S. was fitted by double exponential functions, and no change in time courses of EPSCs was observed following LTD when comparing those time constants, so that a contribution of a

EPSPs was occluded with the LTD induced by conjunctive PF and CF stimulation. It is therefore likely that both forms of persistent depression share common signal transduction mechanisms. Previous reports have shown that CRF is concentrated in the olivo-cerebellar CF system (Cummings et al. 1994b), and CRF expression appears to be upregulated by experimental manipulations to increase the activity of inferior olive neurons. These manipulations included sustained optokinetic stimulation in the rabbit (Barmack and Errico 1993) and Harmaline treatment in the rat (Cummings et al. 1994a). Thus, it is likely that CRF released from CF terminals in certain conditions may play a role in the induction of LTD.

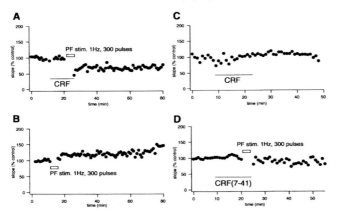

Fig. 1A-D. A Corticotropin-releasing factor (CRF) induced persistent depression of parallel fiber (PF)-EPSPs of cerebellar Purkinje cells (PCs). For this and the following graphs, the initial slope of PF-EPSPs was plotted against time. Each data point repre-sents the slope of the averaged PF-EPSP of five consecutive stimuli expressed as percentage of the baseline before CRF application. Test PF stimulation was repeated at a rate of 0.2 Hz. The *horizontal bar* indicates the period of CRF (0.5 μM) application. During the last 5 min of CRF application, PFs were repetitively stimulated at 1Hz. **B** Repetitive PF stimulation alone without CRF application did not induce depression. **C** CRF application alone without PF stimulation did not induce depression. **D** An inactive analogue of CRF, CRF(7-14) (1μM), with repetitive PF stimulation did not induce depression

Acknowledgments. This work has been partly supported by the Special Postdoctoral Researchers Program, CREST (Core Research for Evolutional Science and Technology) of Japan Science and Technology Corporation (JST).

References

Barmack NH, Errico P (1993) Optokinetically evoked expression of corticotropin-releasing factor in inferior olivary neurons of rabbits. J Neurosci 13:4647-4659
Bishop GA (1990) Neuromodulatory effects of corticotropin releasing factor on cerebellar Purkinje cells: an in vivo study in the cat. Neuroscience 39:251-257

Cummings SL, Hinds D, Young III, WS (1994a) Corticotropin-releasing factor mRNA increases in the inferior olivary complex during harmaline-induced tremor. Brain Res 660:199-208

Cummings SL, Young III WS, King JS (1994b) Early evelopment of cerebellar afferent systems that contain corticotropin-releasing factor. J Comp Neurol 350: 534-549

Gabr RW, Gladfelter WE, Birkle DL, et al (1994) In vivo microdialysis of corticotropin releasing factor (CRF): calcium dependence of depolarization-induced neurosecretion of CRF. Neurosci lett 169:63-67

Potter E, Sutton S, Donaldson C, et al (1994) Distribution of corticotropin-releasing factor receptor mRNA expression in the rat brain and pituitary. Proc Natl Acad Sci USA 91:8777-8781 epifluorescence microscopy. External solutions were quickly changed with a laminar flow jump.

The frequency of miniature excitatory postsynaptic currents (mEPSCs) was low in normal Krebs' solution, but markedly increased in a high K^+ solution. mEPSCs were obviously caused by the activation of nicotinic acetylcholine receptors (nAChR), since they were reversed at around 0 mV and suppressed by 20 μM d-tubocurarine. In half of neurons studied, the amplitude of mEPSCs was gradually increased during high K^+ treatment. The potentiation of mEPSCs amplitude lasted for 15 to 60 min. In the cells, where the potentiation of mEPSCs occurred, the amplitude of membrane currents (I_{nAChR}) induced by acetylcholine (ACh: 10 μM) or DMPP was also potentiated after high K^+ treatment (140 %). On the other hand, the amplitude of I_{nAChR} was not significantly changed in neurons that showed no potentiation of mEPSCs. The potentiation of mEPSCs was accompanied by a concomitant rise in $[Ca^{2+}]_i$ in high K^+ solution. Potentiation of mEPSCs and a rise in $[Ca^{2+}]_i$ were inhibited by BAPTA loaded into cells through a patch pipette. After conditioning application of 100 μM DMPP or ACh for 30 to 60 s, I_{nAChR} induced by a short pulse of ACh (10 μM: 0.3 to 0.6 s) was potentiated to 140 % of the control for 30 to 60 min. This suggests that activation of the nAChR itself potentiates mEPSCs. I_{nAChR} produced by a higher concentration of ACh was less potentiated by treatment with high K^+ or nicotinic agonist. Internal perfusion of KN-62 (30 μM), a specific inhibitor of Ca^{2+}/calmodulin dependent protein kinase II (Ca^{2+}/CaMK II) (Tokumitsu et al. 1990), but not KN-04, an inactive isomer, abolished the potentiation of I_{nAChR} as well as mEPSCs. The results suggest the involvement of Ca^{2+}/CaMK II dependent phosphorylation process in the mechanism.

These results suggest that Ca^{2+} that enters through nAChR channels causes the activation of CaMKII that enhances the sensitivity of nAChR itself to ACh directory or indirectly through protein phosphorylation.

References

Tokumitsu H, Chijiwa T, Hagiwara M, et al (1990) KN-62, 1-[N,O-bis(5-isoquinolinesulfonyl)-N-methyl-L-tyrosyl]-4-phenylpoperazine, a specific inhibitor of Ca^{2+}/calmodulin-dependent protein kinase II. J Biol Chem 265:4315-4320

Introductory Review: Synaptic Development, Structural Modulation, and Gene Expression

M. KANO

Department of Physiology, Kanazawa University School of Medicine, Kanazawa 920-8640, Japan

Key words. Synapse formation, Neuromuscular junction (NMJ), Acetylcholine (ACh), End-plate, Nicotinic acetylcholin receptor (nAChR), Agrin, MuSK, Rapsin, α-Dystroglycan, Knockout mouse, Synapse elimination, Activity, Retrograde messenger, Competition, Proteolysis

1 Introduction

Individual neurons of the brain have complex structures that consist of dendritic trees, somata, and branching axons. Neurons are interconnected through numerous synapses to form complex neural networks. Proper functions of the nervous system are dependent on the formation of specific interconnections between neurons. Thus, developmental neuroscience aims to elucidate how neurons acquire specific identities and how synaptic connections are established and maintained.

The number of genes available to an animal (about 10^5 in mammals) is far from sufficient to specify all the synaptic connections in the brain (about 10^{14}). The development of the nervous system therefore requires the mechanisms by which specific sets of genes are activated in a combinatorial manner at different times during development. Such mechanisms involve both "intrinsic" and "extrinsic" influences. The intrinsic influences that originate from the embryo include intercellular signals mediated by diffusible factors and surface molecules. The extrinsic influences include nutritive factors from the external environment, in addition to the experiences and learning of the animals, which are reflected in changes in neural activity. Establishment of mature pattern of functional synaptic connections is considered to proceed in the following six distinct stages:

(1) Induction of uniform population of neural precursor cells from ectoderm.
(2) Differentiation of neural precursor cells into glial cells and immature neurons.
(3) Migration of immature neurons to their final positions.
(4) Extension of axons from neurons into the vicinity of their eventual targets.

distributions were not significantly different between the two strains (p>0.05, *chi*-square test). In contrast, the frequency distributions between the two strains during the third postnatal week (P15-P21) were markedly different (Fig. 2A, bottom panel) (p<0.01, *chi*-square test). This indicates that a higher percentage of the PCs were multiply innervated by CFs in the mGluR1 mutant than in the wild-type mice. These results suggest that regression of multiple CF innervation initially occurs normally but that the process is specifically impaired during the third postnatal week in the mGluR1 mutant mice.

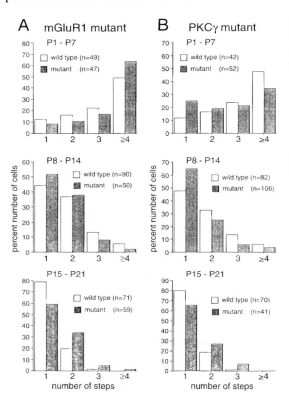

Fig. 2. Postnatal development of CF innervation in mGluR1 **(A)** and PKCγ **(B)** mutant mice. Summary histograms showing number of discrete steps of CF-EPSCs of the wild type (*open columns*) and mGluR1 or PKCγ mutant (*hatched columns*) for the first (P1-P7) (*upper panel*), second (P8-P14) (*middle panel*), and third (P15-P21) (*lower panel*) postnatal week.

2.4 Other Features of the mGluR1 Mutant Cerebellum

We did not find any abnormality in the kinetics, pharmacology, and short-term synaptic plasticity of mGluR1 mutant CF-EPSCs. In addition, the kinetics, pharmacology, and short-term synaptic plasticity of PF-EPSCs were largely normal. These results suggest that CF and PF synapses onto PCs are functional in mGluR1 mutant mice. Moreover, no obvious morphological abnormality was found in the cerebellum of mGluR1 mutant mice. Furthermore, morphology of

PCs immunostained with antibody against spot 35/calbindin appeared normal in the mutant mice. Morphology of individual synapses in the molecular layer were examined by electron microscopy. In both wild-type and mGluR1 mutant mice, the majority of these structures were asymmetric and consisted of presynaptic axonal terminals and postsynaptic dendrites, which are typical of PF-PC synapses. The density of PF-PC synapses in the mGluR1 mutant was not different from that of the wild-type. These observations indicate that gross development of the cerebellar anatomy, differentiation of the PC morphology, and formation of the PF-PC synapses are normal in mGluR1 mutant mice (Kano et al. 1997).

3 Analysis of PKCγ Mutant Mice

3.1 PKCγ Mutant Mice Have Persistent Multiple CF Innervation

PKCγ mutant mice are viable and fertile, but exhibit mild ataxia. The number of CFs innervating individual PCs were estimated with the same electrophysiological technique as used in the analysis of the mGluR1 mutant mice. In most PCs (33 t of 34) from P22-P35 wild-type mice, large EPSCs were elicited in an all-or-none fashion as the stimulus intensity was gradually increased (Fig. 1B, left panel). In contrast, in 40.9% (29 of 71) of the PKCγ mutant PCs, the EPSCs appeared in two to three discrete steps as the stimulus intensity was increased above the threshold (Fig. 1B, left panel). In only 8.8% of the wild-type PCs, the EPSCs had more than one discrete step. The frequency distribution of PCs in terms of the number of discrete EPSC steps showed a significant difference between the wild type and mutant mice (Fig. 2B, right panel) ($P<0.001$, *chi*-square test). This indicates that most PCs in adult wild-type mice are innervated by a single CF, while about one-third of the PCs are innervated by more than one CF in the adult mutant mice.

3.2 Early Postnatal Development of CF Innervation

During the first postnatal week (P1-P7), the majority of PCs were multiply innervated by CFs in both wild-type and PKCγ mutant mice (Fig. 2B, top panel). The frequency distribution of the number of CF-EPSC steps per PC in the PKCγ mutant mice was not significantly different from that in the wild-type mice ($P>0.05$, *chi*-square test). During the second postnatal week (P8-P14), the percentage of PCs with multiple CF-EPSC steps decreased markedly in both wild-type and PKCγ mutant mice, although about 50% of PCs were still multiply

1 Introduction

Long-term potentiation (LTP) is an activity-dependent strengthening of synaptic efficacy that may contribute to memory storage processes in the brain. An analysis of the LTP decay rates in the dentate gyrus showed a classification of LTP into three distinct groups (Abraham et al. 1993). These have been termed "LTP1" with a time constant of about 2 h, "LTP2" with a decay constant of about 4 days, and "LTP3" with an average decay constant of about 23 days. There is good evidence that the synthesis of new macromolecules is important for the development of the longer-lasting forms of LTP, but not for LTP1. Altered gene expression is important for LTP3 but not LTP2, and this is supported by the fact that the mRNA inhibitor actinomycin D does not affect the early stages of LTP2 maintenance (Nyugen et al. 1994). Immediate early genes (IEGs) are rapidly induced in neurons by synaptic activity and are hypothesized to be the macromolecular response required for long-term plasticity. In the LTP model, a number of IEG transcription factors show rapid and transient increases in expression following stimulation, including c-Jun, Jun-B, Fos-related genes, and Zif/268. Of these, Zif/268 has been most reliably induced in correspondence with the induction of LTP. A high-affinity DNA-binding site for zif/268 has been characterized; however, little is known about the target genes regulated by this transcription factor. We have used differential cloning techniques to identify genes that are rapidly induced by neural activity. In contrast to the earlier focus on transcription factors, newly characterized IEGs encode molecules that directly modify the function of neurons. In this review, we summarize recent progress on these "effector" IEG proteins, which functionally induce synaptic reorganizations that underlie activity-dependent neural plasticity.

2 "Effector" Immediate Early Genes

2.1 Cox-2 (Cyclooxygenase-2)

The rate-limiting step in the formation of prostanoids is the conversion of arachidonic acid to prostaglandin H_2 by cyclooxygenase. Two forms of cyclooxygenase have been characterized: a ubiquitously expressed form (Cox-1) and a second form (Cox-2) inducible by various factors including mitogens, hormones, serum, and cytokines. We have cloned cox-2 gene from rat brain (Yamagata et al. 1993). This gene is expressed throughout the forebrain and is enriched in the cortex and hippocampus. Cox-2 mRNA is rapidly and transiently induced by seizures or NMDA-dependent synaptic activity. Basal expression is also regulated by natural synaptic activity. Furthermore, both basal and induced expression of cox-2 are inhibited by glucocorticoids. To elucidate the functional role of Cox-2 in brain diseases, we first examined the effect of brain ischemia on the expression of Cox-2. Transient forebrain ischemia by occluding bilateral carotid arteries in gerbils rapidly induced cox-2 mRNA in the hippocampus, and this increase in mRNA expression persisted for at least 24 h (Ohtuski et al. 1996). Next, we tested the relationship between LPS-induced fever and Cox-2. Injection of LPS markedly induced cox-2 mRNA in two different constituents of the brain: neurons and vascular endothelial cells. As LPS-induced fever was suppressed by pretreatment with a Cox-2-specific inhibitor, we proposed a molecular model, in which induced Cox-2 produces PGE_2 that binds to PGE receptors in the anterior hypothalamus and elevates the body temperature (Cao et al. 1995, 1996, 1997). Cox-2 is a therapeutic target of the widely used nonsteroidal antiinflammatory drugs (NSAIDs). Therefore, if Cox-2 induced by brain diseases is involved in neuronal damage, selective cyclooxygenase-2 inhibitors may be a new therapeutical approach.

2.2 Rheb (Ras Homologue Enriched in Brain)

Rheb is an inducible small GTP-binding protein (Yamagata et al. 1994) The predicted amino acid sequence of Rheb (184 amino acids) is most closely homologous to yeast Ras1 and human Rap2. The putative GTP-binding regions are highly conserved. Recombinant Rheb binds GTP and exhibits intrinsic GTPase activity. Like H-Ras, the carboxyl terminal sequence encodes a CAAX box that is predicted to signal posttranslational farnesylation and to target Rheb to specific membranes. Rheb interacts with and regulates Raf-1 kinase (Yee and Worley 1997). In contrast to H-Ras, however, the interaction of Rheb with Raf-1 is potentiated by growth factors in combination with agents that increase cAMP levels. Protein kinase A-dependent phosphorylation of the regulatory domain of Raf-1 reciprocally potentiates its interaction with Rheb and decreases its interaction with H-Ras. Rheb may function in concert with H-Ras to integrate cAMP and growth factor signaling.

Rheb mRNA is rapidly and transiently induced in hippocampal granule cells by seizures and by NMDA-dependent synaptic activity in the LTP paradigm. It is expressed at comparatively high levels in normal adult cortex as well as a number of peripheral tissues. In the developing brain, rheb mRNA is expressed at relatively high levels in E19 cortical plate. Recently, locus of human rheb gene was determined on chromosome 7q36, using fluorescence in situ hybridization (Muzuki et al. 1996). Considering the chromosomal localization as well as the potential function of this protein, it is important to examine whether rheb is involved in the etiopathogenesis of holoprosencephaly type 3, whose susceptible locus is also linked to 7q36.

to be potential targets of NRSF/REST (Table 1). NRSF/REST has been proposed to be a master negative regulator to restrict expression of neuronal genes to neurons. We examined the contribution of NRSF/REST to NR1 neuronal specificity.

We found that the NRSE/RE1 of NR1 gene works as a silencer in HeLa epitheloid cells. Binding activity of NRSF/REST to NR1 NRSE/RE1 was detected in HeLa cells, but not in neuronally differentiated P19 cells. Overexpression of NRSF/REST reduced NR1 promoter activity in neuronal P19 cells. These findings indicate that NR1 gene is one of the target genes of NRSF/REST.

To determine if NRSF/REST restricts NR1 expression to neurons, correlation of NRSF/REST binding activity with NR1 mRNA level was examined in four nonneuronal cell lines and two neuronal cell lines. The examined nonneuronal cells were HeLa epitheloid cells, PTA lymphoid cells, C6 glioma cells, and P19 embryonal carcinoma cells. The neuronal cells include PC12 pheochromocytoma cells and retinoic acid-treated, neuronally differentiated P19 cells. NRSF/REST binding activity was detected and NR1 mRNA was undetectable in two nonneuronal HeLa and PTA cell lines. In two neuronal PC12 and retinoic acid-treated P19 cells, binding activity of NRSF/REST was not detected and NR1 mRNA was highly expressed. Surprisingly, binding activity of NRSF/REST was not detected in C6 and undifferentiated P19 cells. In these two nonneuronal cells, expression of NR1 mRNA was found, although the expression level was lower than that of neuronal cells.

NRSF/REST activity was not seen in both nonneuronal cell lines (C6 and P19) and neuronal cell lines (PC12 and retinoic acid-treated P19). However, the mRNA levels of NR1 in these nonneuronal cells were lower than that of the neuronal cell lines. Therefore, these results imply that the absence of NRSF/REST is necessary but not sufficient for NR1 expression in neuronal cells. In addition, other neuron-specific transactivating mechanism(s) are necessary for NR1 expression of neuronal cells.

3 The GC-Rich Element Bound by Ubiquitous Transcription Factors, SP1 and MAZ, Is Responsible for NR1 Promoter Activity in Neuronally Dfferentiated P19 Cells

To characterize an additional mechanism for NR1 expression, we determined the NR1 promoter element responsible for NR1 expression in a neuronal differentiation system of P19 cell lines. P19 is a mouse embryonal carcinoma cell line. Retinoic acid treatment converts P19 cells to neurons. Under our culture conditions, more than 99% of cells shows the neuronal phenotype morphologically and by neuronal marker staining. The expression level of NR1 mRNA is enhanced during the differentiation. The binding activity of

NRSF/REST is undetectable in both undifferentiated and differentiated P19 cells. Therefore, the difference in NR1 mRNA level is likely to be due to the more elaborate mechanism(s).

The varied length of NR1 promoter was linked to luciferase reporter gene. The promoter activity was examined in both P19 and neuronally differentiated P19 cells. The promoter activity was stronger in retinoic acid-treated P19 cells, as mRNA level is higher in retinoic acid treated P19 cells. The 5' deletion of the NR1 promoter down to -303 did not affect the promoter activity. However, the deletion from -303 to -276 severely reduced the promoter activity. These results imply that the region from -303 to -276 is responsible for expression of NR1 gene in neuronally differentiated P19 cells. The sequence of the element is

5'-GAAGCGGGGGCGGTGGGAGGGGTAGAA-3'
3'-CTTCGCCCCCGCCACCCTCCCCATCTT-5'

It is GC rich. This type of element is the so-called GC-rich region.

Binding factor(s) to the NR1 GC-rich region were examined by gel shift assay. Two bands were detected in undifferentiated P19 cells. Two similar bands were detected in retinoic acid-treated P19 cells. Antibody to MAZ inhibited the lower binding activity, and antibody to SP1 inhibited the upper binding activity. These results indicate that SP1 and MAZ bind to the NR1 GC-rich region. The binding activities of SP1 and MAZ are not remarkably changed after neuronal differentiation of P19 cells. It has been reported that SP1 and MAZ are ubiquitous transcription factors and bind to GC-rich regions of various genes (Kadonaga et al. 1987; Bossone et al. 1992).

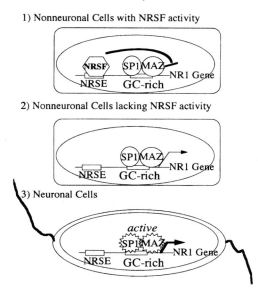

Fig. 1. Three types of regulation of NR1 gene. In nonneuronal cells with NRSF/REST activity (**1**), NRSF/REST blocks the NR1 expression. In nonneuronal cells lacking NRSF and with ubiquitous MAZ (**2**), SP1 bind to the GC-rich region and drive the basal level of transcription. In neuronal cells (**3**), active MAZ and SP1 bind to the GC-rich region and induce high-level expression of the NR1 gene

2 Glutamate-Induced Changes of the Neuron in Hippocampal Slice Culture

2.1 Glutamate Induces Rapid Nuclear Changes

By using a VEC-DIC microscope (Terakawa et al. 1991), we observed hippocampal neurons in a cultured slice, and assessed the changes in morphological appearance at a high magnification in real time. Neurons showed a nucleus of rather smooth appearance when superfused with glutamate-free medium (Fig.1a). Addition of glutamate to the medium at a concentration of 100 μM induced a granulation in the nucleus of the CA1 neurons within 6 min (Fig.1b). In 20 min, the nucleus became completely round in shape, and the soma swelled significantly. The granular changes of the nucleus induced by glutamate were reversible. When glutamate was removed from the medium after a 5- to 15-min application, the nucleus recovered from the granulation and returned to the initial smoothness in 10 min (data not shown).

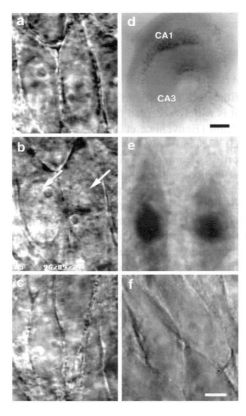

Fig. 1a-f. Selective degeneration of CA1 neurons in a cultured slice. Left column: CA1 neurons were observed before (**a**) and 6 min after (**b**) exposure to 100 μM glutamate. Nuclei (*arrow*) lost their smooth appearance and increased granularity. CA3 neurons were also observed in the same slice 23 min after exposure to glutamate (**c**). Nuclei in CA3 region showed no change. Right column: A hippocampal slice was stained with 0.4% trypan blue 48 h after brief (15 min) exposure to 1 mM glutamate (**d**), and the CA1 (**e**) and CA3 (**f**) regions were observed at high magnification. Trypan blue-positive neurons were found only in the CA1 region. *Bar* in f represents 5 μm for **a**, **b**, **c**, **e**, **f**. *bar* in d, 500 μm. The organotypic slice culture of hippocampus was prepared as described earlier (Stoppini et al. 1991; Yamamoto et al. 1996).

2.2 Brief Exposure to Glutamate Induces Delayed Neuronal Death

We examined delayed effects of the brief glutamate exposure by the trypan blue exclusion technique. Forty-eight hours after a 15-min glutamate application on neurons, the number of cells which could be stained with 0.4% trypan blue was clearly increased in CA1 but not in CA3 regions (Fig. 1d). Each neuron in CA1 showed condensation of the nucleus and formation of blebs. Both the cytoplasm and the nucleus were strongly stained with trypan blue (Fig. 1e). These trypan blue-positive cells were not found 12 h after the brief exposure to glutamate (data not shown).

2.3 Glutamate Induces Morphological Changes Selectively in CA1 Neurons.

Contrary to the neurons in the CA1 subregion of the hippocampus, those in the CA3 region showed neither significant rapid changes (Fig. 1c) nor delayed neuronal death (Fig. 1f) upon glutamate challenge.

3 The Effect of Laser Photolysis of Caged Glutamate on Dissociated Neurons

3.1 Local Uncaging of Glutamate Produces Two Types of Ca^{2+} Transients

To stimulate neurons exclusively at their dendritic region, we developed a laser-spot photochemical activation technique with caged L-glutamate. Using a laser spot of 1 μm in diameter, glutamate was applied in a small puff to a single nerve fiber of isolated CA1 neurons. To examine the responses of the neuron to such pinpoint stimulation, we measured the intracellular Ca^{2+} concentration with a fluorescent indicator, fluo-3. The amplitude of the Ca^{2+} transient in the soma was different from fiber to fiber stimulated, while no significant increase in Ca^{2+} level was observed when the laser pulses were applied without use of the caged compound. A small Ca^{2+} transient was observed with a train of laser pulses (pulse width, 5 ns; 4 Hz for 1 s) applied on one nerve fiber, whereas a larger one was seen on another (Fig. 2a,b). The Ca^{2+} transients became longer and persistent when laser pulses were applied for a longer period (20 Hz for 12 s).

Ikeda J, Terakawa S, Murota S, et al. (1996) Nuclear disintegration as a leading step of glutamate excitotoxicity in brain neurons. J Neurosci Res. 43:613-622

Friedman EJ, Haddad GG (1993) Major differences in Ca^{2+}_i response to anoxia between neonatal and adult rat CA1 neurons: role of Ca^{2+}_o and Na^{2+}_o. J. Neurosci 13:63-72

Park CK, Nehls DG, Graham DL, et al. (1988) The glutamate antagonist MK-801 reduces focal ischemic brain damage in the rat. Ann Neurol 24:543-551

Stoppini L, Buchs PA, Muller D (1991) A simple method for organotypic cultures of nervous tissue. JNeurosciences Methods 37:173-182

Terakawa S, Fan JH, Kumakura K, et al. (1991) Quantitative analysis of exocytosis directly visualized in living chromaffin cells. Neurosci Lett 123:82-86

Wahlestedt C, Golanov E, Yamamoto S, et al (1993) Antisense oligodeoxynucleotides to NMDA-R1 receptor channel protect cortical neurons from excitotoxicity and reduce focal ischaemic infarctions. Nature 363:260-263

Yamamoto S, Sakurai T, Terakawa S, et al. (1996) Glutamate induces rapid nuclear changes in rat organotypic hippocampal culture: video enhanced contrast study. Adv Neurotrauma Res 8:41-45

Role of Adhesion Molecule L1 in Neurite Outgrowth and Functional Synapse Formation

Z-G. ZHONG, S. YOKOYAMA, M. NODA, and H. HIGASHIDA

Department of Biophysical Genetics, Kanazawa University Graduate School of Medicine, Kanazawa 920-8640, Japan

Key words. Extracellular matrix, Protein, Differentiation, NG108-15 neuroblastoma cells

The neuronal cell adhesion molecule L1, which belongs to the extracellular matrix superfamily, is an integral membrane glycoprotein which plays an important role in neuron-neuron adhesion (Rathjen and Schachner 1984), development of growth cones and neurites (Bixby et al. 1988), and migration of cerebellar neurons (Miura et al. 1992). Recently, it has been reported that L1 is strongly induced by osteogenic protein 1 in neuroblastoma x glioma hybrid NG108-15 cells (Perides et al. 1993), Osteogenic protein 1 elicits no significant changes in morphology of NG108-15 cells, which is not consistent with the proposed role of L1. To resolve this discrepancy, we examined the effect of overexpression of L1 protein by L1 cDNA transient transfection on neuronal function in NG108-15 cells. We used lipofectamine, with which 50% of the transfected NG108-15 cell population can express.

L1 overexpression in NG108-15 cells by cDNA transfection did not produce any specific change in neurite outgrowth without dibutyryl cAMP, in agreement with the previous report (Perides et al. 1993). However, surprisingly, we have found that the effect of L1 on neuronal morphology is enhanced under the condition when the cells were differentiated by treatment of dibutyryl cyclic AMP (Zhong et al. 1997). Statistically meaningful differences were obtained in five different parameters such as neurite outgrowth, percentage of cells with neurites, percentage cells with branched neurites, number of neurites per cell, and average length of neurites.

Next, we determined whether the expression of L1 in NG108-15 cells alters functional synapse formation and how L1-enhanced neurite outgrowth contributes to synapse formation in vitro. We found that overexpression of L1 in NG108-15 cells increased percentage of muscle cells with miniature endplate potentials, indicating that L1 facilitates functional synapse formation. These results suggest that L1-overexpression in NG108-15 cells *enhances* promotion of the neurite outgrowth with braches initiated by dibutyryl cAMP treatment (Fig. 1). Our results appear to show that L1 has no direct role at the transcriptional level in initiating neural differentiation, but can promote the differentiation process initiated by cAMP, at least in steps well downstream from transcription.

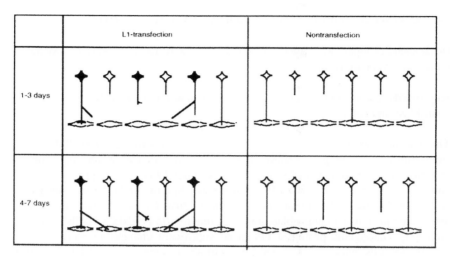

Fig. 1. Scheme shows L1-facilitated synaptic connection with myotubes by increased number of elongated neurites with enhanced branching in the presence of dibutyryl cyclic AMP. The rate of synapse formation by L1-transfection is not different during the early phase of coculture (1-3 days), but was higher during the late phase (4-5 days), than for nontransfected cells. *Filled stars* indicate NG108-15 cells with L1 overexpressed; *open stars* indicate control NG108-15 cells; *spindle* represents a myotube

References

Bixby JL, Lilien J, Reichardt LF (1988) Identification of the major proteins that promote neuronal process outgrowth on Schwann cells in vitro. J Cell Biol 107:353-361

Miura M, Asou H, Kobayashi M, et al (1992) Functional expression of a full-length cDNA coding for rat neural cell adhesion molecule L1 mediates homophilic intercellular adhesion and migration of cerebellar neurons. J Biol Chem 267:10752-10758

Perides G, Hu G, Rueger DC, Charness ME (1993) Osteogenic protein 1 regulates L1 and neural cell adhesion molecule gene expression in neural cells. J Biol Chem 268:25197-25205

Rathjen FG, Schachner M (1984) Immunocytological and biochemical characterization of a new neuronal cell surface component (L1 antigen) which is involved in cell adhesion. EMBO J 3:1-10

Zhong Z-G, Yokoyama S, Noda M, et al (1997) Overexpression of adhesion molecule L1 in NG108-15 neuroblastoma x glioma hybrid cells enhances dibutyryl cyclic AMP-induced neurite outgrowth and functional synapse formation with myotubes. J Neurochem 68:2291-2299

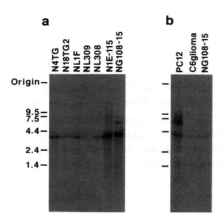

Fig. 2a,b. Blot hybridization analysis of RNA from neuroblastoma- and glioma-derived cells. **a** The samples analyzed were 15 μg of total RNA from N4TG neuroblastoma cells, N18TG2 neuroblastoma cells, and N18TG2 x B82 fibroblast hybrid NL1F, NL309 and NL308 cells (Minna et al. 1972; Higashida et al. 1990), 4.9 μg of poly(A)$^+$ RNA from N1E-115 neuroblastoma cells (Higashida and Brown 1987), and 4.5 μg of poly (A)$^+$ RNA from NG108-15 cells. **b** Fifteen micrograms of total RNA from PC12 rat pheochromocytoma cells, C6 rat glioma cells, and NG108-15 cells was analyzed. The blots were hybridized with ^{32}P-labeled mouse (**a**) or rat (**b**) B$_2$ BKR cDNA as described previously (Hashii et al. 1996). Hybridization signals were visualized on an imaging analyzer (BAS1000; Fuji, Japan). An RNA ladder (Bethesda Research Laboratories) was used as size markers (in kilobases).

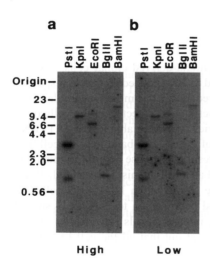

Fig. 3a,b. Southern blot analysis of rat genomic DNA: 2 μg of genomic DNA from Wistar rat kidney was digested with indicated restriction enzymes. Restriction fragments were separated by electrophoresis in 0.75% agarose gels. The filters were hybridized at 42□C with ^{32}P-labeled rat B$_2$ BKR cDNA under conditions of high stringency in buffer containing 50% formamide (**a**) and low stringency in buffer containing 30% formamide (**b**), and then washed at 65 °C in 0.1 x SSC containing 0.1% SDS (**b**) and at 50 °C in 1 x SSC containing 0.1% SDS (**a**). Hybridization signals were visualized on an imaging analyzer (BAS1000). Note that no additional band was observed under low stringency condition. The size markers used were *Hin*dIII-digested λ DNA (in kiliobases).

The data presented here indicate that the B$_2$ BKR transcripts initially observed in NG108-15 cells are expressed in various CNS regions and other neuronal and glial cell lines. It is conceivable that these transcripts are expressed from the same gene as for the B$_2$ BKR of the smooth muscle type. Our conclusion agrees with the earlier work by Borkowski et al. (1995), which demonstrated that in mice targeted disruption of the B$_2$ BKR gene for the smooth muscle type abolished BK responses not only in uterine smooth muscle but also in superior cervical ganglion neurons. It seems most likely that the pharmacological diversity of the B$_2$ BKR is generated by posttranscriptional process(es).

Acknowledgments.   This work was supported by grants from Kowa Life Science Foundation and from the Ministry of Education. Science, Sports and Culture of Japan.

References

Borkowski JA, Ransom RW, Seabrook GR, et al (1995) Targeted disruption of a B$_2$ bradykinin receptor gene in mice eliminates bradykinin action in smooth muscle and neurons. J Biol Chem 270:13706-13710

Braas KM, Manning DC, Perry DC et al (1988) Bradykinin analogues: differential agonist and antagonist activities suggesting multiple receptors. Br J Pharmacol 94:3-5

Hashii M, Nakashima S, Yokoyama S, et al (1996) Bradykinin B$_2$ receptor-induced and inositol tetrakisphosphate-evoked Ca^{2+} entry is sensitive to a protein tyrosine phosphorylation inhibitor in *ras*-transformrd NIH/3T3 fibroblasts. Biochem J 319:649-656

Higashida H, Brown DA (1987) Bradykinin inhibits potassium (M) currents in N1E-115 neuroblastoma cells: responses resemble those in NG108-15 neuroblastoma x glioma hybrid cells. FEBS Lett 220:302-306

Higashida H, Okano Y, Hoshi N, et al (1990) Bradykinin induces inositol 1,4,5-trisphosphate-dependent hyperpolarization in K$^+$ M-current-deficient hybrid NL308 cells: comparison with NG108-15 neuroblastoma x glioma hybrid cells. Glia 3:1-12

Higashida H, Hashii M, Yokoyama S, et al (1996) Bradykinin B$_2$ receptors and signal transduction analyzed in NG108-15 neuroblastoma x glioma hybrid cells, B$_2$ receptor-transformed CHO cells and *ras*-transformed NIH/3T3 fibroblasts. Prog Brain Res 113:215-230

Llona I, Vavrek R, Stewart J, et al (1987) Identification of pre- and postsynaptic bradykinin receptor sites in the vas deferens: evidence for different structural prerequisites. J Pharmacol Exp Ther 241:608-614

McEachern AE, Shelton ER, Bhakta S, et al (1991) Expression cloning of a rat B$_2$ bradykinin receptor. Proc Natl Acad Sci USA 88:7724-7728

Minna J, Glazer D, Nirenberg M (1972) Nat New Biol 235:225-231

Nardone J, Gerald C, Rimawi L, et al (1994) Identification of a B$_2$ bradykinin receptor expressed by PC12 pheochromocytoma cells. Proc Natl Acad Sci USA 91:4412-4416

Noda M, Ishizaka N, Yokoyama S, et al (1996) Inositol trisphosphate/Ca^{2+} as messengers of bradykinin B$_2$ and muscarinic acetylcholine m1-m4 receptors in neuroblastoma-derived hybrid cells. J Lipid Mediat Cell Signal 14:175-185

Pelvin R, Owen PJ (1988) Multiple B$_2$ bradykinin receptors in mammalian tissues. Trends Pharmacol Sci 9:387-389

Rifo J, Pourrat M, Vavrek RJ, et al (1987) Bradykinin receptor antagonists used to characterize the heterogeneity of bradykinin-induced responses in rat vas deferens. Eur J Pharmacol 142:305-312

Taketo M, Yokoyama S, Rochelle J, et al (1995) Mouse B$_2$ bradykinin receptor gene maps to distal chromosome 12. Genomics 27:222-223

Walker K, Perkins M, Dray A (1995) Kinins and kinin receptors in the nervous system. Neurochem Int 26:1-16

2 Sympathetic Ganglia

2.1 Postsynaptic Modulation of Ganglionic Transmission

Adrenaline (Ad) and noradrenaline (NA) hyperpolarize amphibian sympathetic neurons by activation of a K^+ conductance acting on postsynaptic α_2- receptors (Kobayashi and Libet 1970; Smith 1994). A depolarizing response has also been reported to be produced by catecholamines (Koketsu and Akasu 1986). A slow depolarization was produced by acetylcholine (ACh) via muscarinic receptors in sympathetic ganglia. ACh also caused a slow hyperpolarizing response via muscarinic receptors. Luteinizing hormone-releasing hormone (LHRH), the transmitter of the late slow EPSP, depolarizes neurons in amphibian sympathetic ganglia by suppressing the M current (Jones 1985). Substance P, the transmitter of the slow EPSP in inferior mesenteric ganglia of the guineapig, produces a long-lasting depolarization (Dun and Minota 1981; Konishi et al. 1979). Other transmitters, such as 5-HT, ATP, and GABA, also caused depolarizations in sympathetic ganglia (Smith 1994) (Table 1A). The sensitivity of nicotinic ACh receptors in postganglionic neurons is modulated by various transmitters (Koketsu and Akasu 1986).

2.2 Presynaptic Modulation of Cholinergic Transmission

Release of ACh in bullfrog sympathetic ganglia can also be controlled by activation of presynaptic receptors with agonists such as Ad and perhaps also by cholinergic autoreceptors (Kuba and Kumamoto 1990). In mammalian sympathetic ganglia, the presynaptic inhibitory effect of NA is quite strong and may account for the overall depressant effect on ganglionic transmission (Christ and Nishi 1971; Keast et al. 1990). Adrenoceptor agonists produce a short-term inhibition and a long-term potentiation (l.t.p.) of ACh release in bullfrog sympathetic ganglia. These effects have been carefully characterized (Briggs 1995; Jänig 1995; Kuba and Kumamoto 1990). Ad applied to sympathetic neurons for about 10 min caused an l.t.p. of the amplitude and quantal content of the fast EPSP, which lasted for several hours after its removal. The l.t.p. of ACh release is mediated via a cyclic AMP-coupled α-adrenoceptor, which may serve to increase the basal level free Ca^{2+} in the presynaptic terminal. Horn and McAfee (1980) showed that Ca^{2+} spikes in mammalian sympathetic ganglia could be suppressed by catecholamines. Conditioning stimulation (50 Hz for 10 s) the preganglionic nerve also causes an l.t.p. (use-dependent l.t.p.) of the fast EPSP (Brown and McAfee 1982). Other transmitters including GABA, 5-HT, muscarine, and histamine produce the presynaptic modulation (Table 1B).

Table 1. Pre- and postsynaptic modulation in sympathetic ganglia

---

A. Postsynaptic modulation

Depolarization	*Hyperpolarization*
Acetylcholine	Muscarine
Nicotinic	Adrenaline
Muscarinic	Noradrenaline
Adrenaline	Dopamine
5-Hydroxytryptamine	Enkephalin
γ-Aminobutyric acid	
Luteinizing hormone releasing hormone	
Substance P	
Adenosine triphosphate	
Prostaglandin E_1	

B. Presynaptic modulation

Inhibition	*Facilitation(l.t.p.)*
Acetylcholine	Adrenaline
Muscarinic	
Adrenaline	
γ -Aminobutyric acid	
Dopamine	
5-Hydroxytryptamine	
Histamine	
Adenosine triphosphate	
Adenosine	
Luteinizing hormone releasing hormone	
Substance P	
Enkephalin	
Vasoactive intestinal peptide	
Cholecystokinin octapeptide	
Vasopressin	
Prostaglandin E_1	

---

l.t.p.,long-term potentiation.

(Surprenant and Crist 1988). The amplitudes of the fast EPSP and slow EPSP were presynaptically reduced by histamine via H_2 receptors (Mihara 1993) (Table 2B).

4 Parasympathetic Ganglia

4.1 Postsynaptic Modulation of Parasympathetic Ganglia

Suzuki and Volle (1979) first reported that NA produced a depolarizing potential associated with decreased input resistance in a majority of neurons in the hamster submandibular ganglion. In cat vesical pelvic ganglia (VPG), NA produced not only a hyperpolarization via α_2-adrenoceptors but also a depolarization via α_1-adrenoceptors (Akasu and Nishimura 1995; Shinnick-Gallagher et al. 1983). Neurons in rabbit VPG exhibit mainly the hyperpolarizing response to NA (Akasu and Nishimura 1995). Many transmitters also produce postsynaptic modulation in parasympathetic ganglia (Table 3A).

4.2 Presynaptic Modulation of Synaptic Transmission in Parasympathetic Ganglia

Neurotransmitters and biogenic substances presynaptically inhibit or facilitate cholinergic transmission by modulating the release of ACh from preganglionic nerve terminals in parasympathetic ganglia. Exogenous NA inhibits cholinergic transmission in vesical pelvic neurons (Akasu and Nishimura 1995; de Groat and Booth 1980; Griffith et al. 1979). The activation of α_2-adrenoceptors reduces the evoked and spontaneous release of ACh from preganglionic nerve terminals. VPG may be the site for integration of sympathetic and parasympathetic inputs to the urinary bladder (de Groat and Booth 1980). Sympathetic nerve activity and exogenous adrenergic agonists can modulate cholinergic transmission at the VPG (de Groat and Booth 1980; Keast et al. 1990). The presynaptic inhibition of cholinergic transmission termed adrenergic 'heterosynaptic' modulation may be an important process that controls the function of the urinary bladder. Opioid peptides released endogenously from preganglionic nerves also mediate the 'heterosynaptic' inhibition of cholinergic transmission via δ-opioid receptor in cat VPG (de Groat and Kawatani 1989). 5-HT has been shown to produce a presynaptic inhibition of cholinergic transmission in ciliary ganglia of the cat (Tatsumi and Katayama 1987) and rabbit VPG (Akasu and Nishimura 1995) via 5-HT_{1A} receptors. Other substances that produce presynaptic modulation of parasympathetic cholinergic transmission are summarized in Table 3B.

Table 3. Pre- and postsynaptic modulation of parasympathetic ganglia

---

A.Postsynaptic modulation

Depolarization	*Hyperpolarization*
Acetylcholine	Acetylcholine
Nicotinic	Muscarinic
Muscarinic	Noradrenaline (α_2)
Adrenaline	γ -Aminobutyric acid (GABA$_A$)
Noradrenaline (α_1)	Adenosine(A$_1$)
5-Hydroxytryptamine (5-HT$_3$)	Dopamine
γ-Aminobutyric acid (GABA$_A$)	Enkephalin
Substance P	
Adenosine triphosphate (P$_2$)	
Enkephalin	
Galanin	
Endothelin-1 (ET$_B$)	
Vasoactive intestinal peptide	

B.Presynaptic modulation

Inhibition	*Facilitation*
Noradrenaline (α_1, α_2)	5-Hydroxytryptamine (?)
5-Hydroxytryptamine (5-HT$_{1A}$)	
Adenosinetriphosphate (P$_2$)	
Adenosine	
Enkephalin (δ)	
Endothelin-1	

---

5 Synaptic Modulation of Sympathetic and Parasympathetic Preganglionic Neurons

5.1 Sympathetic Preganglionic Neurons (SPNs)

Among neurotransmitters that are present in the spinal cord, the effects of amines, such as NA, Ad and 5-HT, and peptides (substance P, angiotensin II, vasopressin) have been studied in in vitro (Inokuchi et al. 1990; Lewis et al. 1993; Ma and Dun 1986; Miyazaki et al. 1989, 1998; Nishi 1990; Pickering et al. 1994). Modulatory effects of glutamate and GABA, through activation of metabotropic and GABA$_B$

3 Electrophysiological Characteristics of SCN Neurons

SCN neurons fire action potentials at a higher frequency during the subjective day than during the subjective night, with the mean population firing frequency peaking near CT6 (Groos and Hendriks 1982; Green and Gillette 1982; Shibata et al. 1982). CT indicates Circadian Time, with lights-on being designated as CT0. During the subjective day, action potential firing frequencies range from less than 1 Hz to 14 Hz, compared to the subjective night, when SCN neurons fire with frequencies of 0.5 to 3 Hz. The prevalence of fast-firing neurons during the day is responsible for the increased mean population firing frequency observed in SCN slices (Gillette 1991). During the subjective night the mean population firing frequency decreases primarily due to a reduction of the fast-firing cells. The majority of SCN cells fire spontaneous action potentials with amplitudes ranging from 50 to 86 mV, with a mean of 63 mV (Jiang et al. 1997). The action potentials are usually preceded by a prepotential, followed by an after-hyperpolarization, and are unaffected by a combination of the AMPA receptor antagonist CNQX (10 μM), the NMDA receptor antagonist APV (50 μM), and the GABA$_A$ receptor antagonist bicuculline (30 μM). Injection of depolarizing current induced firing in quiescent cells, and a larger current increased the firing frequency. The maximum depolarization-induced firing frequency varied from cell to cell with a mean of 19 Hz. Long-term recordings of individual neurons show a similar pattern with the firing rates gradually decreasing from CT6 to CT18 and slowly increasing from CT18 to CT6 (Groos and Hendriks 1982; Cahill and Menaker 1989; Kim and Dudek 1991). These data indicate that SCN neurons are spontaneously active as the result of intrinsic ionic mechanisms, not the afferent synaptic excitatory drive.

4 Regulation of Circadian Rhythms by Glutamate

The SCN receives environmental light information from neurons in the retina that project directly to the SCN via the retinohypothalamic tract (RHT) and use glutamate as a neurotransmitter (Cahill and Menaker 1989; Kim and Dudek 1991; Jiang et al. 1997). The circadian clock regulates its own sensitivity to the resetting properties of glutamate (Gillette 1997). Glutamate will phase delay the circadian clock when applied early in the night (CT 13-15), while later in the night (CT 18-20) glutamate will produce a phase advance (Ding et al. 1994; Gillette 1991; Gillette 1986). Interestingly, the glutamate phase response curve overlaps that for light (Gillette 1991). SCN neurons express N-methyl-D-aspartate (NMDA), (±)-α-amino-3-hydroxy-5-methylisoxazole-4-propionic acid (AMPA), and metabotropic glutamate receptors (Card and Moore 1991; Scott and Rusak 1996). A role for activation of the NMDA receptors in phase-shifting actions of glutamate and light has been demonstrated (Ding et al. 1994; Rea et al. 1993).

Bath application of AMPA or NMDA induced inward currents in SCN neurons recorded from brain slices. Currents induced by AMPA (10 µM) and NMDA (10 µM) were reduced 85% to 90% by CNQX (10 µM) and APV (50 µM), respectively (Jiang et al. 1997). The AMPA current amplitude was linearly related to the membrane potential and reversed its polarity at approximately 9 mV. In contrast, the NMDA-induced currents attained a maximum amplitude between -50 and -70 mV; both depolarization and hyperpolarization reduced the current. When recorded in the same cell, the onset of the NMDA-induced current was slower than that of the AMPA current.

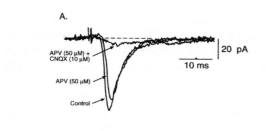

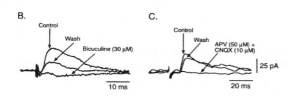

Fig. 1A-C. Characteristics of EPSCs and IPSPs recorded in the SCN. **A** The amplitude of EPSCs elicited by a stimulus (25 V, 0.2 ms) to the optic chiasm were reduced 13% by APV (50 µM) and 90.5% (3) by a combination of APV (50 µM) and CNQX (10 µM). The *broken line* indicates the baseline current. **B** Focal stimulation of the SCN evoked an outward synaptic current, which was blocked by bicuculline (30 µM) and recovered slowly. **C** A long latency IPSC was evoked by stimulation of the contralateral optic nerve and was blocked by a combination of CNQX (10 µM) and APV (50 µM) or bicuculline, suggesting the involvement of inhibitory interneurons within the SCN. Each trace in B and C was an average of 5 sweeps. Holding potential for A, B, and C was -60 mV.

Spontaneously released glutamate-activated inward currents were CNQX sensitive and identified as fast EPSCs. The amplitude and frequency of spontaneous EPSCs varied considerably from cell to cell but were relatively constant in individual neurons. The amplitude of the spontaneous EPSCs ranged from 5 to 102 pA and the frequency ranged from 0.4 to 58 Hz. The mean EPSP rise time (from 10% to 90% of peak amplitude) was 1.3 ± 0.2 ms, and the decay was well fit with a single exponential function with a mean time constant of 3.1 ± 0.27 ms (Jiang et al. 1997).

Stimulation of the optic nerve or the optic chiasm evokes inward currents in the majority of SCN cells. Stimulation of the ipsilateral optic nerve was more effective than that of the contralateral optic nerve in evoking an inward current (Fig. 1). Supramaximal stimulation evoked EPSCs of 3 to 192 pA in amplitude and from 5 to 40 ms in duration (Jiang et al. 1997). The majority of the evoked EPSCs had a monophasic waveform, with a mean rise time of 1.2 ms and a decay time constant of 3.4 ms. An estimate of the minimum number of presynaptic fibers converging onto a single neuron was made by examining the discrete increments and inflections on the synaptic currents as the stimulus voltage was slowly raised from threshold to supramaximal. From these data, approximately five RHT axons were estimated to converge onto each SCN neuron.

The EPSCs were reversibly reduced 13% by APV (50 µM) alone and 95% by a combination of APV (50 µM) and CNQX (10 µM), identifying them as the fast EPSCs mediated by both AMPA and NMDA receptors. The evoked EPSC amplitude was enhanced by hyperpolarization from -60 mV and reduced and reversed in its polarity by depolarization. The reversal potential of the EPSC was estimated to be 3 mV (Mayer and Westbrook 1987). The duration of the EPSC became longer when the cell was depolarized. When CNQX (10 µM) was present, the residual EPSC was depressed by hyperpolarization. The maximum inward current appeared between -50 and -30 mV and was reduced to near zero at -120 mV, suggesting a Mg^{2+} block of the NMDA receptor-mediated current (Mayer and Westbrook 1987). Optic nerve stimulation in some neurons evoked a biphasic current (inward followed by an outward current). Bicuculline (30 µM) or a high extracellular calcium concentration (40 mM) blocked only the outward current, but CNQX (10 µM) and APV (50 µM) blocked both the inward and outward currents, suggesting that the outward current responses result from the activation of an inhibitory interneuron. The conduction velocity of the retinohypothalamic fibers has been estimated to be 0.22 to 0.54 m/s, consistent with a view that fine unmyelinated axons from a group of small retinal ganglion cells conduct the nerve pulses (Moore et al. 1995; Jiang et al. 1997). Therefore, AMPA and NMDA receptors are responsible for mediating both spontaneous and optic nerve-stimulated release of glutamate from the RHT.

5 Regulation of Circadian Rhythms by GABA

Several lines of evidence demonstrate that GABA plays a key role in the activity of the SCN pacemaker. The majority of neurons and about 50% of the synaptic boutons in the SCN are GABA immunoreactive, and GABA-mediated inhibitory synaptic potentials (IPSP) can be recorded from virtually all SCN cells which receive excitatory retinal inputs (Jiang et al. 1995; Moore and Speh 1993; Decavel and Van den Pol 1990; Kim and Dudek 1992). The $GABA_A$ receptor antagonist bicuculline blocks the light-induced phase delay of locomotor rhythms. Moreover, a single injection of the benzodiazepine analog, triazolam, produces a permanent phase shift in behavioral and endocrine rhythms by interacting with the benzodiazepine-binding site located on the $GABA_A$ receptor (Ralph and Menaker 1985). The specific $GABA_B$ receptor agonist baclofen prevents light-induced phase advances, phase delays, and the photic induction of c-*fos* in SCN (Ralph and Menaker 1989; Colwell et al. 1993). Baclofen also modifies optic nerve-evoked field potentials and inhibits the release of glutamate from terminals of the RHT (Jiang et al. 1995; Gannon et al. 1995).

Spontaneous outward currents were blocked by bicuculline but not by strychnine, suggesting they were $GABA_A$ receptor-mediated fast IPSCs induced by spontaneously released GABA. The isolated IPSCs had mean rise times of 2.2 ms and decays with a mean time constant of 7.1 ms. Both the rise and decay of IPSC were significantly slower than those of EPSC ($P<0.05$). The amplitude of IPSCs was enhanced by depolarization and reduced and reversed in polarity by hyperpolarizations to and beyond -75 mV. TTX (1 μM) suppressed the large-amplitude but not the small-amplitude spontaneous IPSCs in a minority of SCN neurons and had no effect in the remaining cells. These results suggested that most spontaneous IPSCs and EPSCs were miniature synaptic currents caused by transmitter release independent of action potentials.

Stimulation of the ipsilateral or contralateral SCN produced a biphasic response consisting of an EPSC followed by a bicuculline-sensitive IPSC (Fig. 1). When the bathing solution contained CNQX (10 μM) and APV (50 μM), focal stimulation of the SCN evoked an outward current that reversed polarity near -70 mV. This current was always blocked by bicuculline (30 μM) and is therefore a $GABA_A$ receptor-mediated IPSC. The outward currents were nullified when the cell was hyperpolarized to about -75 mV and the currents were blocked by bicuculline (3-30 μM), suggesting they were IPSCs induced by spontaneous GABA release.

Baclofen (1 μM), a $GABA_B$ agonist, reversibly reduced the amplitude of EPSCs evoked by optic nerve stimulation. 2-Hydroxysaclofen (300 μM), a $GABA_B$ antagonist, reduced the inhibition produced by baclofen. Baclofen reduced the frequency but not the amplitude of the spontaneous miniature EPSCs. In a subset of SCN neurons, baclofen (1 μM) also induced an outward current and a conductance increase. Baclofen had no effect on either evoked or spontaneous IPSCs or on currents activated by pulse application of glutamate. From these data we conclude that baclofen can act directly on SCN neurons to activate a K^+ conductance and inhibit the release of glutamate from terminals of the RHT, and suggest the in vivo physiological actions of baclofen are mediated by these two mechanisms (Jiang et al. 1995).

6 Coupling of Melatonin Receptors to K^+ Channels in SCN Neurons

Melatonin is an indole amine neurohormone produced in the pineal gland which shows a circadian rhythm in blood levels. Melatonin levels are low during the day, begin to increase just before dusk, and rise through most of the night to peak just before dawn (Skene et al. 1990). Melatonin acting in the SCN can entrain the circadian rhythms of rats maintained in constant darkness or light (Cassone et al. 1986; Redman et al. 1983). Melatonin can also phase shift the daily rhythmic

activity of SCN neurons *in vitro* (McArthur et al. 1991). The SCN and the pars
tuberalis are the primary sites of high-affinity melatonin binding in the brains of
both rodents and humans (Reppert et al. 1988; Vanecek et al. 1987). Therefore,
the SCN is believed to be a major site of action for the modulation of circadian
function by melatonin (Cassone et al. 1986). The melatonin receptor is coupled to
a pertussis toxin-sensitive G protein and mediates the circadian and reproductive
functions of melatonin (Reppert et al. 1994; Ebisawa et al. 1994).

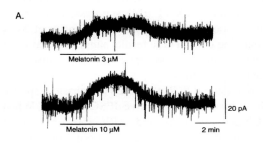

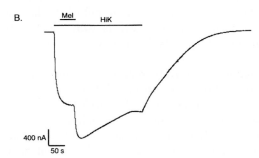

Fig. 2A,B. Coupling of melatonin receptors to K⁺ channels. **A** Melatonin (3 and 10 µM) activated an outward current in an SCN neuron voltage-clamped at -60 mV. The melatonin was applied during the subjective daytime between CT9 and CT12. **B** Oocytes expressed the *Xenopus* melatonin receptor and Kir3.1 and Kir3.2 subunits were voltage clamped at -60 mV. A HiK Ringer's (80 mM K⁺) and 1 µM melatonin were applied as indicated by *bars*. The melatonin-activated current demonstrated a functional coupling between melatonin receptors and G-protein activated K⁺ channels.

Melatonin applied between CT9 and CT12 activates an outward current in 35%
of SCN cells with an estimated EC_{50} of 1.2 µM (Fig. 2). Melatonin (10 µM)
induced an outward current in 23% of cells when applied between CT 6-9 and
25% of neurons recorded at CT12-15. The mean amplitudes of the melatonin
current at CT6-CT9 and at CT12-CT15 were not significantly different from that
of CT9-CT12. The melatonin-activated current was augmented by depolarization,
reduced by hyperpolarization, and reversed its polarity near the K⁺ equilibrium
potential. Ba^{2+} (1 mM) blocked 80-100% of melatonin-induced current while Cs^{2+}
(3 mM) reduced the melatonin-induced current by 20-50%. In contrast to a
$GABA_B$ agonist, melatonin (1-30 µM) had no effect on either the amplitude or
frequency of spontaneous or evoked EPSCs or IPSCs. These data suggest that
melatonin is directly activating a G-protein coupled K⁺ channel.

Kir3 channels are G-protein activated K⁺ channels that are directly stimulated
by G-protein βγ subunits giving inwardly rectifying K⁺ currents (Takao et al.
1994; Kofuji et al. 1995). Four Kir3 subunits (Kir3.1-4) have been cloned, and

heteromeric channel assembly is probably required for physiological K^+ channel function (Kofuji et al. 1995). We hypothesized that the K^+ currents activated by melatonin in SCN neurons were Kir3 channels. The *Xenopus* oocyte expression system was used to determine whether melatonin receptors could activate Kir3 channels. Melatonin-activated currents were observed in oocytes expressing a *Xenopus* melatonin receptor, Kir3.1 and Kir3.2 cRNAs (Fig. 2). G-proteins of the $G_i\alpha$ family are sensitive to ADP ribosylation by pertussis toxin (Ptx), which results in the formation of inactive Gi heterotrimers. In *Xenopus* melatonin receptor-expressing oocytes, Ptx (0.5 μg/ml) blocked melatonin-stimulated currents in a time-dependent manner. Thus, melatonin receptor coupling to Kir3 channels is Ptx sensitive and mediated by G_i-proteins. Therefore, melatonin receptors activate heteromeric Kir3.1/Kir3.2 channels when expressed in oocytes and may couple to similar K^+ channels expressed in SCN neurons.

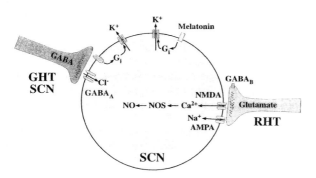

Fig. 3. Proposed model of synaptic transmission in the suprachiasmatic nucleus.

Summary and Conclusion. These data suggest a picture of the synaptic mechanism that sends environmental information to the circadian clock (Fig. 3). Afferent terminals from the retina terminate in the SCN. The release of glutamate acts on AMPA receptors to mediate fast synaptic excitatory transmission. At hyperpolarized membrane potentials, NMDA receptors will be blocked by extracellular Mg^{2+}. However, at more depolarized potentials glutamate will open NMDA receptors, which allow Ca^{2+} to enter the cell, possibly activating nitric oxide synthase and increasing the levels of nitric oxide. GABA is the primary inhibitory neurotransmitter in the SCN and acts on both $GABA_A$ and $GABA_B$ receptors. Acting on bicuculline-sensitive $GABA_A$, receptors produced an inhibitory potential and hyperpolarized the membrane potential. $GABA_A$ receptors are located on both presynaptic and postsynaptic membranes. The presynaptic receptors inhibit the release of transmitter from glutamatergic terminals, possibly by inhibiting voltage-gated calcium channels. Postsynaptically the $GABA_B$ receptors can activate potassium channels. The hormone melatonin, the receptors of which are coupled to G_i-type G-proteins, can also activate K^+ channels. The synaptic regulation of the circadian clock is a complex process involving receptors coupled to both iontropic and G-protein coupled receptors. The elements of the

Ralph MR, Menaker M (1985) Bicuculline blocks circadian phase delays but not advances. Brain Res 325:362-365

Ralph MR, Menaker M (1989) GABA regulation of circadian responses to light. I. Involvement of $GABA_A$-benzodiazepine and $GABA_B$ receptors. J Neurosci 9:2858-2865

Rea MA, Buckley B, Lutton LM (1993) Local administration of EAA antagonists blocks light-induced phase shifts and c-*fos* expression in hamster SCN. Am J Physiol Regul Integr Comp Physiol 34:R1191-R1198

Redman J, Armstrong S, Ng KT (1983) Free-running activity rhythms in the rat: entrainment by melatonin. Science 219:1089-1091

Reppert SM, Schwartz WJ (1984) The suprachiasmatic nuclei of the fetal rat: characterization of a functional circadian clock using 14C-labeled deoxyglucose. J Neurosci 4:1677-1682

Reppert SM, Weaver DR, Rivkees SA, et al (1988) Putative melatonin receptors in a human biological clock. Science 242:78-81

Reppert SM, Weaver DR, Ebisawa T (1994) Cloning and characterization of a mammalian melatonin receptor that mediates reproductive and circadian responses. Neuron 13:1177-1185

Scott G, Rusak B (1996) Activation of hamster suprachiasmatic neurons *in vitro* via metabotropic glutamate receptors. Neuroscience 71:533-541

Shibata S, Oomura Y, Kita H, Hattori K. (1982) Circadian rhythmic changes of neuronal activity in the suprachiasmatic nucleus of the rat hypothalamic slice. Brain Res 247:154-158

Skene DJ, Vivien-Roels B, Sparks DL, et al (1990) Daily variation in the concentration of melatonin and 5-methoxytryptophol in the human pineal gland. Brain Res 528:170-174

Takao K, Yoshii M, Kanda A, et al (1994) A region of the muscarinic-gated atrial K^+ channel critical for activation by G protein $\beta\gamma$ subunits. Neuron 13:747-755

Van den Pol AN (1980) The hypothalamic suprachiasmatic nucleus of rat: intrinsic anatomy. J Comp Neurol 191:661-702

Van den Pol AN, Tsujimoto KL (1985) Neurotransmitters of the hypothalamic suprachiasmatic nucleus: Immunocytochemical analysis of 25 neuronal antigens. Neuroscience 15:1049-1086

Van den Pol AN (1986) Gamma-aminobutyrate, gastrin releasing peptide, serotonin, somatostatin, and vasopressin: ultrastructural immunocytochemical localization in presynaptic axons in the suprachiasmatic nucleus. Neuroscience 17:643-659

Vanecek J, Pavlik A, Illnerova H. (1987) Hypothalamic melatonin receptor sites revealed by autoradiography. Brain Res 435:359-362

Welsh DK, Logothetis DE, Meister M, et al (1995) Individual neurons dissociated from rat suprachiasmatic nucleus express independently phased circadian firing rhythms. Neuron 14:697-706

Fast and Slow Synaptic Responses to Dorsal Root Simulation in Dorsal Commissural Nucleus Neurons of Rat Sacral Spinal Cord In Vitro

H. INOKUCHI[1], Y. LU[1,2], and H. HIGASHI[1]

[1]Department of Physiology, Kurume University School of Medicine
67 Asahi-machi, Kurume 830-0011
[2]Department of Anatomy, The Fourth Military Medical University
Xi'an, 710032 People's Republic of China

Key words. Fast EPSP(C), Fast IPSP(C), Slow EPSP, Glutamate, AMPA receptor, NMDA receptor, GABAergic, Glycinergic, Substance P, NK$_1$ receptor, Aδ fiber activation, C fiber activation, Dorsal commissural nucleus, Sacral spinal cord, Viscerovisceral reflexes

Summary. Activation of Aδ or Aδ plus C sensory afferents exhibits three kinds of synaptic response in the DCN neurons of the sacral spinal cord. Fast excitatory synaptic responses were mediated through the activation of AMPA and/or NMDA receptors by glutamate released from both Aδ and C sensory afferents. On the other hand, fast inhibitory responses were mediated by activation of GABA$_A$ or glycine receptors via GABAergic or glycinergic interneurons activated by sensory afferents. In addition, two kinds of slow excitatory synaptic response were elicited by repetitive stimulation of sensory afferents: one was mediated via NMDA receptor by activation of Aδ afferents; and the other was mediated by activation of NK$_1$ receptors by substance P released apparently from C fiber sensory afferents. Thus, DCN neurons integrate direct excitatory and indirect excitatory and inhibitory fast inputs and modulate these fast inputs by slow excitatory responses.

1 Introduction

Neurons of the dorsal commissural nucleus (DCN) in the lumbosacral spinal cord (L6-S1) receive somatic and visceral afferent inputs from pelvic viscera and respond to noxious and non-noxious stimulation (Birder and deGroat 1992; Birder et al. 1991; Honda 1985; Ness and Gebhart 1987). These neurons are likely to be

205930 (20 µM), a 5HT₃ antagonist, completely blocked the CNQX and APV-resistant fast EPSPs (n=1). Further analysis is required to determine whether 5HT is a transmitter for this fast EPSP.

The fast IPSC(P)s that occurred alone (Fig. 1Ba) or in combination with EPSC(P)s (Fig. 1Bb,c) were blocked by bicuculline (20 µM) and/or by strychnine (2 µM). Fast IPSCs that occurred alone were reversed in polarity at membrane potential of -75 mV (n=2) and disappeared in the presence of CNQX. The results suggest that these were indirect GABAergic and/or glycinergic inputs activated through Aδ afferent activation.

3 Slow Synaptic Responses

Slow synaptic responses to dorsal root stimulation were observed in only 1 neuron of 28 neurons of transverse slices. However, slow postsynaptic responses were readily detected in longitudinal slices although they were rarely evoked by single stimuli.

Repetitive stimulation (10-20 stimuli at 10 or 20 Hz) of Aδ fibers (0.6-4.2V, 0.4 ms) in normal Krebs solution elicited, in addition to fast synaptic responses, slow depolarizations (slow EPSPs) in 45% of neurons in which fast EPSPs were recorded (n=38; Fig. 2A). Slow EPSPs were evoked without generating any fast synaptic responses in only one neuron. In the rest of the neurons in which slow EPSPs were not elicited by Aδ fiber activation, increasing stimulus intensity of C fiber activation evoked slow EPSPs in only one case.

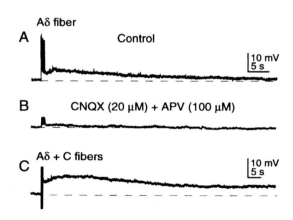

Fig. 2A-C. Slow postsynaptic responses. Aδ fiber activation (1 V, 0.4 ms, 20 stimuli at 20 Hz) evoked a slow EPSP (A) that was eliminated by CNQX and APV (B). Increasing the strength of stimulation (10 V, 0.4 ms, 20 stimuli at 20 Hz) to activate C fibers (C) yielded a larger and more prolonged slow response. Membrane potential was -58 mV.

Aδ fiber-mediated slow EPSPs evoked in normal Krebs solution ranged from 30 s to 2.5 min in total duration and 2.5 mV to 8 mV in amplitude. Time to peak of the slow EPSPs was also variable from cell to cell, ranging from 1.2 s to 13 s.

A role for NMDA-receptor-mediated slow synaptic responses has been identified in synaptic plasticity of dorsal horn neurons in neonatal rat (Miller and

Woolf 1996). APV (100 μM) applied alone completely blocked the slow EPSPs without affecting fast EPSPs in 2 of 3 neurons tested. The APV-sensitive slow EPSPs were 30 to 40 s in total duration and 3 and 4 mV in amplitude. Further, in 50% of neurons with slow EPSPs, CNQX (20 μM) in combination with APV (100 μM) completely blocked Aδ fiber-mediated slow EPSPs (Fig. 2B); however, increasing stimulus intensity to activate C fiber inputs elicited an additional slow EPSP (Fig. 2C). The slow EPSP was associated with increased membrane resistance. This result implies that primary afferent C fibers terminate on neurons in the DCN and elicit slow EPSPs by releasing a substance other than glutamate. Aδ plus C fiber-mediated slow EPSPs had similar time courses to those of Aδ-evoked slow EPSPs; i.e., time to peak was 4.8 to 20 s, total duration 1 to 2 min, and amplitude 2.5 to 7.5 mV (see Table 1).

Table 1. Properties of slow responses evoked by brief trains of stimuli at different stimulus strength (Mean ± SD, n=9 neurons)

	Time to peak (s)	1/2 Decay time (s)	Total duration (s)	Amplitude (mV)
Control (Aδ fibers)	3.7 ± 1.9	15.7 ± 10.8	67.5 ± 33.8	5.2 ± 2.4
CNQX+APV (Aδ+C fibers)	6.5 ± 5.4	23.0 ± 11.6	82.9 ± 30.6	5.1 ± 1.9

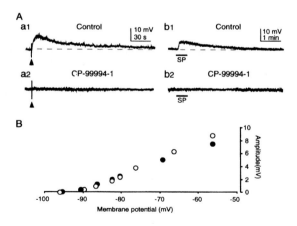

Fig. 3A,B. Effect of NK$_1$ antagonist on slow EPSP and substance P (*SP*) induced membrane depolarization. Records were obtained in the presence of CNQX (20 μM) and APV (100 μM) at membrane potential of -65 mV. Stimuli (10 V, 0.4 ms, 5 pulses at 20 Hz) were applied to dorsal root. **B.** Plot of peak amplitude of SP-induced response at different membrane potentials.

According to immunohistochemical studies, substance P (SP) is very likely to be a transmitter candidate for the generation of slow EPSPs in the DCN. SP and its antagonists, therefore, were examined in the presence of both CNQX 20 μM and 100 μM APV. SP (8.6 μM, applied for 30 s) produced a slow membrane depolarization (2-10 mV in amplitude, lasting 30 s to 5 min) with increased neurons form local circuits within the lateral septum (Jakab and Leranth 1995) to mediate the inhibitory postsynaptic potential (IPSP) (Gallagher et al. 1995). The massive projection originated from hippocampal CA1 and CA3 neurons terminates mainly on principal neurons in the lateral septum (Alonso and Frotscher 1989; Jakab and Leranth 1995). These principal neurons also receive afferents from dopaminergic inputs from the ventral tegmental area (VTA), noradrenergic input from locus coeruleus (LC), and serotonergic inputs from dorsal raphe nucleus (DR) (Gall and Moore 1984; Köhler et al. 1982; Lindvall and Stenevi 1978).

It has been known that neurotransmitters are capable of modulating either membrane excitability or synaptic transmission in the mammalian central nervous system (CNS) (Andrade et al. 1986; North 1989). 5-Hydroxytryptamine (5-HT) produces a hyperpolarizing response in dorsolateral septal nucleus (DLSN) neurons (Joëls and Urban 1984; Joëls et al. 1987). However, few electrophysiological studies concerning the role of serotonergic neurons on the pathway between the hippocampus and the lateral septum have been reported because of difficulties of simultaneous recording of electrical signals by using multiple intracellular microelectrodes. Recently, optical recording with voltage-sensitive dye has enabled us to record spatial and temporal spreading of neuronal activities at multiple sites simultaneously (Cohen and Lesher 1986; Grinvald et al. 1988; Kamino 1991; Salzberg 1989). We investigated the role of 5-HT in the hippocampo-septal pathway by using electrophysiological and optical recording techniques.

2 Electrophysiological Studies for the Role of 5-HT in the Lateral Septum

2.1 Hyperpolarization Induced by 5-HT

Intracellular recordings were made from DLSN neurons in transverse slices (500 μm thick) of male Wistar rats (Stevens et al. 1984). Neurons in the DLSN had a resting membrane potential of -70 ± 3 mV ($n=45$) and input resistance of 132 ± 11 MΩ ($n=45$). Bath application of 5-HT (100 μM) caused a hyperpolarizing response with amplitude of 13 ± 4 mV ($n=12$) in DLSN neurons (Fig. 1A). Spontaneous firing of action potentials of DLSN neurons was blocked during the hyperpolarization. The 5-HT hyperpolarization was associated with decrease in input resistance. 8-Hydroxy-di-n-propylamino tetralin (8-OH-DPAT, 3 μM), an agonist for 5-HT$_{1A}$ receptor subtype, produced a hyperpolarizing response associated with decreased membrane resistance (Fig. 1B). CP 93129 (10 μM), a 5-HT$_{1B}$ receptor agonist (100 μM), and L-694,247, a 5-HT$_{1D/1B}$ agonist (100 μM), caused hyperpolarizing responses. The order of agonist potency was 5-HT $\geqq$ 8-

OH-DPAT >>CP 93129 = L-694,247. RS 67333, a 5-HT$_4$ receptor agonist, and 2,5-dimethoxy-4-iodophenylisopropylamine (DOI, 40 μM), a 5-HT$_2$ receptor agonist, caused no hyperpolarizing response in DLSN neurons. NAN-190 (0.5-10 μM), a selective antagonist for the 5-HT$_{1A}$ receptor subtype, strongly reduced the hyperpolarization induced by 5-HT. These results suggest that 5-HT produces hyperpolarization through the 5-HT$_{1A}$ receptor subtype in DLSN neurons.

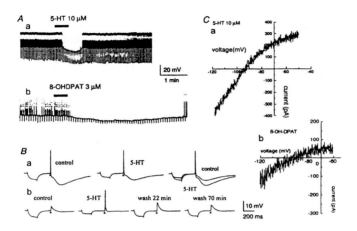

Fig. 1A-C. Effects of 5-hydroxytryptamine and 8-OH-DPAT on the membrane potential (**A**), postsynaptic potentials (**B**), and membrane current (**C**) in DLSN neurons. **A** *Horizontal bars* indicate the period of 5-HT application. **B** Postsynaptic potentials were evoked by stimulation of fimbrial pathway. **C** *Panels a* and *b* show inward rectifier K$^+$ currents produced by 5-HT (10 μM) and 8-OH-DPAT (3 μM), respectively.

2.2 Contribution of the Inward Rectifier K$^+$ Current to 5-HT Hyperpolarization

The hyperpolarization induced by 5-HT was strongly reduced by Ba^{2+} (1 mM) but not by tetraethylammonium (TEA, 3 mM), extracellular Cs$^+$ (1 mM), and glibenclamide (100 μM). It is well known that intracellular Cs$^+$ blocks K$^+$ channels in a nonselective manner (Hille 1992). Injection of Cs$^+$ to DLSN neurons reduced the amplitude of the 5-HT-induced hyperpolarization. Under voltage-clamp condition, 5-HT (1-100 μM) caused an outward current (I$_{5-HT}$) associated with increase in membrane conductance. The reversal potential of I$_{5-HT}$ was −90 ± 3 mV (n=22). I$_{5-HT}$ shows inward rectification at a potential more negative than -70 mV in about 50% of DLSN neurons (Fig. 1C). 8-OH-DPAT also increased the inward rectifier K$^+$ current. Ba^{2+} (100 μM) blocked the inward rectifier type of I$_{5-HT}$. In some neurons, I$_{5-HT}$ showed outward rectifier properties at membrane potentials more positive than -70 mV. Many neurotransmitter receptors have been reported potentials along nerve fibers. The latter (Ca^{2+}-sensitive) component of optical signals spreading to the lateral septum involves synaptic events. The synaptic component of the optical response was enhanced in the presence of (50 μM) 5-HT (Fig. 2A). Field potentials were simultaneously recorded by extracellular microelectrode from the same brain slices for optical recordings. 5-HT (100 μM) increased the amplitude of compound action potentials (Fig. 2B). The facilitation of optical signals was not mimicked by DOI (40 μM). 8-OH-DPAT also enhanced the optical response. These results suggest that the 5-HT-induced facilitation of optical images is mediated, at least in part, by 5-HT_{1A} receptors in DLSN neurons.

4 Proposed Role of Serotonergic Neurons in the Lateral Septum

With optical recording techniques, spatiotemporal spread of neuronal activity was recorded from brain slice the hippocampal CA3 and the lateral septum. Optical responses initially occurred near the stimulus electrode and then spread toward the dorsolateral portion of the lateral septum. The optical signals mediated by excitatory transmission were enhanced in the presence of 5-HT (50 μM). The 5-HT-induced augmentation of signal propagation to the lateral septum is probably due to both the direct facilitation of the EPSP and the depression of the IPSPs. Field potentials simultaneously recorded by extracellular electrode from the same brain slices were also enhanced by 5-HT. Principal neurons in the lateral septum containing GABA receives massive hippocamposeptal afferents containing main axons of CA1 and CA3 hippocampal pyramidal neurons (Alonso and Frotscher 1989; Jakab and Leranth 1995). These septal neurons not only send GABAergic afferents to hypothalamic and amygdaloid areas but also form local inhibitory circuits within the lateral septum to downregulate excitatory synaptic transmission (Jakab and Leranth 1995). The present study showed that 5-HT inhibited the IPSPs in DLSN neurons. The disinhibition of GABAergic interneurons may also lead the facilitation of the excitatory synaptic transmission in the lateral septum. Additionally, the soma membrane of DLSN neurons showed a hyperpolarizing response when 5-HT was applied to the brain slice preparation. During the 5-HT hyperpolarization, spontaneous firing of action potentials was blocked at the postsynaptic membrane of septal neurons. These results suggest that 5-HT enhances signal-to-noise ratio to selectively pass the information of hippocampal CA1 and CA3 neurons to lower brain centers.

References

Alonso JR, Frotscher M (1989) Organization of the septal region in the rat brain: a Golgi/EM study of lateral septal neurons. J Comp Neurol 286:472-487

Andrade R, Malenka RC, Nicoll RA (1986) A G protein couples serotonin and $GABA_B$ receptors to the same channels in hippocampus. Science 234:1261-1265

Cohen LB, Lesher S (1986) Optical monitoring of membrane of multisite optical measurement. In: De Weer PJ, Salzberg BM (eds) Optical methods in cell physiology. Wiley, New York, pp 71-99

Gall C, Moore RY (1984) Distribution of enkephalin, substance P, tyrosine hydroxylase, and 5-hydroxytryptamine immunoreactivity in the septal region of the rat. J Comp Neurol 225:212-227

Gallagher JP, Zheng F, Hasuo H, et al (1995) Activities of neurons within the rat dorsolateral septal nucleus (DLSN). Prog Neurobiol 45:373-395

Grinvald A, Frostig RD, Lieke E, et al (1988) Optical imaging of neuronal activity. Physiol Rev 68:1285-1366

Hasuo H, Shoji S, Gallagher JP, et al (1992) Adenosine inhibits the synaptic potentials in rat septal nucleus neurons mediated through pre- and postsynaptic A_1-adenosine receptors. Neurosci Res 13:281-299

Hille B (1992) Ionic channels of excitable membranes, 2nd edn. Sinauer, Sunderland

Iijima T, Witter MP, Ichikawa M, et al (1996) Entorhinal-hippocampal interactions revealed by real-time imaging. Science 272:1176-1179

Jakab RL, Leranth C (1995) Septum. In: Paxinos G (ed) The rat nervous system, 2nd edn. Academic Press, San Diego, pp 405-442

Joëls M, Urban IJA (1984) Electrophysiological and pharmacological evidence in favor of amino acid neurotransmission in fimbria-fornix fibers innervating the lateral septal complex of rats. Exp Brain Res 54:455-462

Joëls M, Shinnick-Gallagher P, Gallagher JP (1987) Effect of serotonin and serotonin analogues on passive membrane properties of lateral septal neurons in vitro. Brain Res 417:99-107

Kamino K (1991) Optical approaches to ontogeny of electrical activity and related functional organization during early heart development. Physiol Rev 71:53-91

Köhler C, Chan-Palay V, Steinbusch H (1982) The distribution and origin of serotonin-containing fibers in the septal area: a combined immunohistochemical and fluorescent retrograde tracing study in the rat. J Comp Neurol 209:91-111

Kubo Y, Reuveny E, Slesinger PA, et al (1993) Primary structure and functional expression of a rat G-protein-coupled muscarinic potassium channel. Nature 364:802-806

Lindvall O, Stenevi U (1978) Dopamine and noradrenaline neurons projecting to the septal area in the rat. Cell Tissue Res 190:383-407

Nakagami Y, Saito H, Matsuki N (1987) Optical recording of rat entorhino-hippocampal system in organotypic culture. Neurosci Lett 216:211-213

North RA (1989) Twelfth Gaddum memorial lecture. Drug receptors and the inhibition of nerve cells. Br J Pharmacol 98:13-28

Salzburg BM (1989) Optical recording of voltage changes in nerve terminals and in fine neuronal processes. Annu Rev Physiol 51:507-526

Stevens DR, Gallagher JP, Shinnick-Gallagher P (1984) Intracellular recordings from rat dorsolateral septal neurons, in vitro. Brain Res 305:353-356

Sugitani M, Sugai T, Tanifuji M, et al (1994). Optical imaging of the in vitro guinea pig piriform cortex activity using a voltage-sensitive dye. Neurosci Lett 165:215-218

Tanifuji M, Sugiyama T, Murase K (1994) Horizontal propagation of excitation in rat visual cortical slices revealed by optical imaging. Science 266:1057-1059

Muscarinic Modulation of Na⁺ Spike Propagation in the Apical Dendrites of Hippocampal CA1 Pyramidal Neurons

H. TSUBOKAWA[1,2], N. KAWAI[1], and W.N. ROSS[2]

[1]Department of Physiology, New York Medical College Valhalla, NY 10595, USA
[2]Department of Physiology, Jichi Medical School Minamikawachi-machi, Tochigi 329-0498, Japan

Key words. Hippocampus, Slice, Pyramidal neuron, Apical dendrites, Dendritic patch, Fluorescence imaging, Na⁺ spike, Backpropagation, Ca²⁺, Muscarinic receptors, G protein, PKC, K⁺ conductance, I_{AHP}, M-current

1 Introduction

In CA1 pyramidal neurons of hippocampus, action potentials initiated near the soma backpropagate over the dendrites in an activity-dependent manner (Jaffe et al. 1992; Regehr and Tank 1992; Callaway and Ross 1995; Spruston et al. 1995). This backpropagation is thought to be important in regulating some forms of synaptic plasticity (Kullman et al. 1992; Magee and Johnston, 1997; Markram et al. 1997). Using dendritic patch recording and high-speed calcium imaging, we examined a number of factors that could affect dendritic spike propagation. The clearest result was obtained from the muscarinic agonist carbachol (CCh), which made all spikes in a train propagate almost equally into the dendrites, increasing the magnitude and spatial extent of the $[Ca^{2+}]_i$ change resulting from the spikes (Tsubokawa and Ross, 1997).

2 Materials and Methods

Preparation of hippocampal slices and recording procedures have been described previously (Tsubokawa and Ross 1997). The bathing solution was composed of (in mM): 124 NaCl, 2.5 KCl, 2 CaCl₂, 2 MgCl₂, 1.25 NaH₂PO₄, 26 NaHCO₃, 10 glucose, bubbled with a mixture of 95% O₂, 5% CO₂. L-Ascorbic acid (0.4 mM) and myo-inositol (1.5-3 mM) were added to this solution. The pipette solution contained (in mM): 130 K-gluconate, 10 NaCl, 2 Mg-ATP, 0.3 Na-GTP, and 10

N-2-hydroxyethylpiperazine-N'-2-ethansulfonic acid (HEPES), pH adjusted to 7.3 with KOH. Open resistance of the pipette was 7-11 MΩ. Tight seals (>5 GΩ) were made using the "blind" (Blanton et al., 1989) approach or under visual control using a 40x water-immersion lens and video-enhanced DIC optics.

Changes in intracellular calcium concentration were measured using the fluorescent calcium indicator calcium green-1 (200 μM, C-3010, Molecular Probes, Eugene, OR) added to the pipette solution. High-speed images (25-ms frame interval) were recorded using a cooled CCD camera (Lasser-Ross et al. 1991). Changes in calcium concentration are presented as the spatial average of ΔF/F (%), where F is the fluorescence intensity at resting membrane potential (corrected for background autofluorescence) and ΔF is the time dependent change in fluorescence (corrected for bleaching).

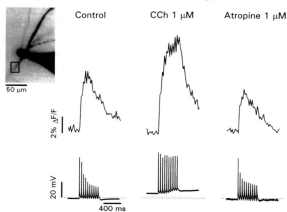

Fig. 1. Carbachol enhances backpropagation of antidromically evoked (20 Hz, 10 pulses) spikes (*lower*) with increase in amplitude and duration of spike-evoked [Ca²⁺]ᵢ changes (*upper*) in the dendrites. Atropine blocks these effects. Inset shows patch electrode and dendrite (250 μm from soma) filled with 200 μM Calcium Green-1. Electrode appears large because of overexposure. Changes in fluorescence averaged over the box close to the electrode were measured.

3 Results & Discussion

Results from a typical experiment are shown in Fig. 1. The image shows the electrode and the dendrites filled with calcium green-1. The recording location was about 250 μm from the soma. In the presence of the glutamate receptor antagonists, 6-cyano-7-nitroquinoxaline-2,3-dione (CNQX, 10 μM) and DL-2-aminophosphonovalerate (APV, 50 μM), and the GABA_A receptor antagonist, bicuculline (10 μM), we recorded a train of action potentials with decreasing amplitudes in response to stimulation in the alveus. At the same time we recorded the fluorescence increase from a rectangular region close to the electrode. In control conditions the increase in fluorescence peaked at the time of the first few, larger action potentials. When 1 μM CCh was added to the bath solution, the amplitude of the later spikes increased. The amplitude of the fluorescence change increased and peaked at the time of the last spike in the train. When 1 μM atropine was added to the solution, both the spike profile and the fluorescence change

References

Isaacson JS, Nicoll RA (1991) Aniracetam reduces glutamate receptor desensitization and slows the decay of fast excitatory synaptic currents in the hippocampus. Proc Natl Acad Sci USA 88:10936-10940

Ito I, Tanabe S, Kohda A, Sugiyama H (1990) Allosteric potentiation of quisqualate receptors by a nootropic drug aniracetam. J Physiol (Lond) 424:533-543

Partin KM, Patneau DK, Mayer ML (1994) Cyclothiazide differentially modulates desensitization of AMPA receptor splice variants. Mol Pharmacol 46:129-138

Patneau DK, Vyklicky L, Mayer ML (1993) Hippocampal neurons exhibit cyclothiazide-sensitive rapidly desensitizing responses to kainate. J Neurosci 13:3496-3509

Petitet F, Blanchard JC, Double A (1995) Effects of non-NMDA receptor modulators on [^{3}H]dopamine release from rat mesencephalic cells in primary cultures. J Neurochem 64:1410-1412

Sekiguchi M, Fleck MW, Mayer ML, et al (1997) A novel allosteric potentiator of AMPA receptors:4-[2-(phenylsulfonylamino)ethylthio]-2,6-difluorophenoxyacetamide. J Neurosci 17:5760-5771

Sekiguchi M, Takeo J, Harada T, et al (1998) Pharmacological detection of AMPA receptor heterogeneity by use of two allosteric potentiators in rat hippocampal cultures. Br J Pharmacol 123:1294-1303

Tang CM, Shi QY, Katchman A, et al (1991) Modulation of the time course of fast EPSCs and glutamate channel kinetics by aniracetam. Science 254:288-290

Tsuzuki K, Takeuchi T, Ozawa S (1992) Agonist- and subunit-dependent potentiation of glutamate receptors by a nootropic drug aniracetam. Mol Brain Res 16:105-110

Yamada K, Tang CM (1993) Benzothiadiazines inhibit rapid glutamate receptor desensitization and enhance glutamatergic synaptic currents. J Neurosci 13:3904-3915

Modulation of Cholinergic Synaptic Transmission by Arachidonic Acid in Bullfrog sympathetic Neurons

S. MINOTA

Department of Basic Medical Science, Kobe City College of Nursing
Gakuen-nishi-machi, Nishi-ku, Kobe 651-2103, Japan

Key words. Arachidonic acid, Cholinergic receptor, Nicotinic receptor, Muscarinic receptor, M-channel, M-current, Fast EPSP, Fast EPSC,Slow EPSP, Slow EPSC, Facilitation, LTP

Arachidonic acid (AA) and its metabolites are the possible second messengers for channel modulations in many cells. AA may modulate ion conductance, either directly (Ordway et al. 1991), or indirectly through lipoxygenase metabolites (Piomelli and Greengard 1990). In hippocampal neurons, AA or its metabolite mediates the somatostatin-induced augmentation of the M-current (Schweitzer et al. 1990). AA reduced nicotine-induced responses in chick ciliary ganglion neurons (Vijayaraghavan et al. 1995). In bullfrog sympathetic neurons, AA increases M-current (Yu 1995), while it reduces the nicotinic receptor-mediated responses (Minota and Watanabe 1997). However, the effects of AA on muscarinic receptor-mediated responses have not been studied in detail. I report here the action of AA on cholinergic synaptic transmission in bullfrog *(Rana catesbeiana)* sympathetic neurons. Single-electrode current- or voltage-clamp recordings were made from B-type neurons.

1 Effects of AA on Nicotinic Response

The fast excitatory postsynaptic potentials (fast EPSPs) and the underlying currents (fast excitatory postsynaptic currents, fast EPSCs) were elicited continuously by single electrical stimulation applied to the preganglionic nerve fibers every 5 s in a Ca^{2+}-deficient solution. AA (0.2-40 µM) reversibly reduced the amplitude of both fast EPSPs and fast EPSCs (Minota and Watanabe 1997). The mean reduction of fast EPSC by 20 µM AA was 43%. In addition to reducing the amplitude of fast EPSCs, AA elicited an outward current in about half of

(Koyano et al. 1985; Minota et al. 1991). In the hippocampus, the released AA from plasma membrane is thought to be a retrograde transmitter for the initiation of LTP (Williams et al. 1989). However, in the case of bullfrog sympathetic ganglia, AA (20 μM) did not significantly affect the pre-LTP (Minota and Watanabe 1997), suggesting that AA may not have a major role in the generation and maintenance of pre-LTP in bullfrog sympathetic neurons.

References

Brown DA, Adams PR (1980) Muscarinic suppression of a novel voltage-sensitive K^+ current in a vertebrate neuron. Nature 283:673-676

Koyano K, Kuba K, Minota S (1985) Long-term potentiation of transmitter release induced by repetitive presynaptic activities in bull-frog sympathetic ganglia. J Physiol (Lond) 359:219-233

Kuba K, Koketsu K (1976) Analysis of the slow excitatory postsynaptic potential in bullfrog sympathetic ganglion cells. Jpn J Physiol 26:651-669

Minota S (1995) Delayed onset and slow time course of the non-M-type muscarinic current in bullfrog sympathetic neurons. Pflugers Arch 429:570-577

Minota S, Kumamoto E, Kitakoga O, et al (1991) Long-term potentiation induced by a sustained rise in the intraterminal Ca^{2+} in bullfrog sympathetic ganglia. J Physiol (Lond) 435:421-438

Minota S, Watanabe S (1997) Inhibitory effects of arachidonic acid on nicotinic transmission in bullfrog sympathetic neurons. J Neurophysiol 78:2396-2401

Ordway RW, Singer JJ, Walsh, JV Jr (1991) Direct regulation of ion channels by fatty acids. Trends Neurosci 14:96-100

Piomelli D, Greengard P (1990) Lipoxygenase metabolites of arachidonic acid in neuronal transmembrane signaling. Trends Pharmacol Sci 11:367-373

Sargent PB (1993) The diversity of neuronal nicotinic acetylcholine receptors. Annu Rev Neurosci 16:403-443

Schweitzer P, Madamba S, Siggins GR (1990) Arachidonic acid metabolites as mediators of somatostatin-induced increase of neuronal M-current. Nature 346:464-467

Vijayaraghavan S, Huang B, Blumenthal EM, et al (1995) Arachidonic acid as a possible negative feedback inhibitor of nicotinic acetylcholine receptors on neurons. J Neurosci 15:3679-3687

Williams JH, Errington ML, Lynch MA, et al (1989) Arachidonic acid induces a long-term activity-dependent enhancement of synaptic transmission in the hippocampus. Nature 341:739-742

Yu SP (1995) Roles of arachidonic acid, lipoxygenases and phosphatases in calcium-dependent modulation of M-current in bullfrog sympathetic neurons. J Physiol (Lond) 487:797-811

Slow Intrinsic Optical Signals in Rat Spinal Cord Slices and Their Modulation by Low-Frequency Stimulation

K. MURASE, H. IKEDA, S. TERAO, and T. ASAI

Department of Human and Artificial Intelligent Systems, Fukui University
3–9–1 Bunkyo, Fukui 910–8507, Japan

Key words. Excitatory amino acids, Intrinsic optical signals, Long-term depression, Opioid, Optical recording, Pain, Primary afferent fibers, Slow synaptic responses, Spinal cord, Spinal dorsal horn, Substance P, Substantia gelatinosa

1 Introduction

The slow synaptic response in the spinal dorsal horn (DH) was originally found by Urban and Randic (1984). Tetanic stimulation of high-threshold primary afferent fibers in the dorsal root (DR) was found to induce a slow depolarization associated with spike and excitatory postsynaptic potential (epsp) discharges in the DH neurons in a slice preparation. Bath application of substance P, the putative pain transmitter/modulator, was shown to mimic the slow synaptic response. Since then, evidence has accumulated that the slow synaptic response lasting for more than 1 mm might take a critical role in nociceptive information transmission (Randic et al.1987; Urban et al.1994; Yoshimura 1996). However, some of the important characteristics including the precise site of cells generating slow synaptic responses and the induction mechanisms are still unclear primarily due to the methodological limitations of intracellular recording techniques.

Activity-dependent intrinsic optical signals (IOSs) in excitable tissue have been used to investigate spatial patterns of neuronal excitation in various mammalian preparations both in vivo (e.g., Grinvald et al.1986, Haglund et al. 1992) and in vitro (e.g., Lipton 1973; Federico et al. 1994). The extent of the IOSs has been shown to correlate well with the extent of neuronal excitation (Grinvald et al.1988, Hopp et al. 1990). Evidence has indicated that IOSs in mammalian hippocampal and cortical slice preparations are attributable to glial cell swelling (Andrew and MacVicar 1994; Holthoff and Witte 1996). An activity-dependent rise of the extracellular K^+ ions triggers a net KCl uptake via the furosemide-sensitive Na^+-

Origin of IOSs

Perfusion of spinal cord slices with an inhibitor of $Na^+–K^+–2Cl^-$ cotransporter, furosemide (5 mM), for 20 min or longer nearly abolished the stimulation-induced IOS. In addition, the IOS was not induced in the low-Cl^- solution.

We tested whether direct stimulation of $Na^+–K^+–2Cl^-$ cotramsporter or cell swelling causes IOSs similar to those induced by electrical stimulation. When the K^+ concentration of the perfusate was increased from 3.1 to 8.0 mM for 1 min and returned to 3.1 mM, the K^+-induced IOSs had the same polarity and a similar amplitude to the stimulus-induced IOS, although the time course was slower. One-minute perfusion of hypotonic solutions (0.98T and 0.95T) evoked IOSs as well.

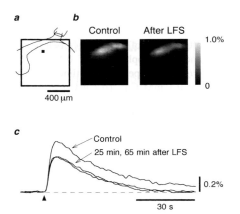

Fig. 2a-c. Spatial distributions and time courses of stimulation-induced slow IOSs before and after 10-min application of low-frequency stimulation. Slow IOSs were elicited by brief high-frequency stimulation (20 Hz for 1 s with current pulses of 0.5 ms duration and 1.2 mA amplitude) of the DR applied every 20 min, and low-frequency stimulation (LFS) (1 Hz for 10 min with current pulses of 0.5-ms duration and 1.5 mA amplitude) was given 5 min after test stimulation. **a** Illustration of a transverse spinal cord slice with an attached DR and a suction pipette. **b** Images of the peak changes in light transmittance 6.1 s after onset of test stimulation 5 min before (*left*) and 65 min after application of LFS (*right*). **c** Superimposed time courses of slow IOSs 5 min before (*Control*) and 25 min and 65 min after application of LFS.

5 LTD of IOSs by LFS

IOSs were evoked by test stimulation (20 Hz for 1 s, 0.9–1.3 mA intensity), which was delivered at every 20 min. The conditioning LFS (1 Hz for 10 min) was applied 5 min after test stimuli. The intensity of the conditioning stimuli was reduced to 0.5 mA and the duration to one tenth (0.05 ms) to activate low-threshold A fibers but not high-threshold C fibers. Five minutes after cessation of 10-min conditioning LFS, amplitudes of IOSs were not decreased and remained unchanged for the next 40–60 min. In four slices tested with low-intensity LFS, amplitudes were 99 ± 1% of the control values at 5 min and 97 ± 3% at 45 min after conditioning LFS.

The intensity of conditioning simulation was raised to a level either identical to that of test stimulation (0.9–1.0 mA) or 1.5 mA to activate C fibers in addition. In an example shown in Fig. 2, an amplitude of IOSs was decreased to 70% of averaged control value at 25 min after LFS and remained depressed for further 40 min. Figure 3 shows time courses of amplitudes of IOSs for six slices that received a 10-min LFS. Although the degree of depression by conditioning LFS varied among slices, the changes lasted for more than 60 min.

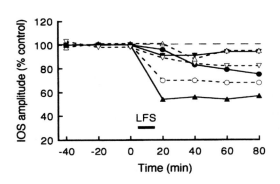

Fig. 3. Time courses of peak amplitude of slow IOSs from six slices that received 10-min conditioning LFS. Peak amplitudes were normalized to averages of those recorded before conditioning LFS. Slow IOSs were elicited by brief high-frequency stimulation (20 Hz for 1 s with current pulses of 0.5-ms duration and 0.9–1.3 mA amplitude) of the DR applied every 20 min. Conditioning LFS (1 Hz for 10 min with current pulses of 0.5-ms duration and 1.5-mA amplitude) was given 5 min after test stimulation, as indicated by *black horizontal bar.*

6 Summary and Discussion

This report has summarized our results regarding the IOS in the mammalian spinal cord slices. The adequate stimulus to induce IOSs in the DH was the tetanic activation of both A and C fibers. The IOS, lasting for several minutes, was most prominent in the SG of the DH and much less in the other parts. The treatment with the Na^+-K^+-$2Cl^-$ cotransport blocker furosemide, or the removal of Cl^- from the perfusate, abolished the IOS, the result suggesting its origin being the cellular swelling caused by an activity-dependent rise of extracellular K^+ (Svoboda and Sykova 1991). Substance P antagonist spantide, glutamate antagonists AP5 and CNQX, and a μ-opioid agonist DAMGO reversibly suppressed IOSs. Thus, IOSs represented at least in part the slow excitatory response that has been known in DH neurons after tetanic activation of nociceptive afferents.

This is the first to report that the IOS can undergo LTD by the conditioning LFS. Electrophysiological studies have shown that C-fiber-evoked responses in spinal DH are inhibited by A-fiber stimulation (Malzack and Wall 1965), and undergo LTD (Sandkühler et al. 1997). However, LFS of afferent A fibers failed to induce LTD of IOSs, whereas LFS of C fibers could induce LTD of IOSs. The

Propagation of Depolarization-Induced Suppression of Inhibition in Cultured Rat Hippocampal Neurons

T. OHNO-SHOSAKU[1], S. SAWADA[2], and C. NAKANISHI[3]

Department of Physiology, Faculty of Medicine, Kanazawa University
13-1 Takara-machi, Kanazawa 920-8640, Japan

Key words. Rat,Hippocampus,DSI,Inhibitory transmission, Depolarization, Cultured neuron,Calcium,IPSC, Autapse, Propagation,Retrograde messenger, Synaptic modulation, Transient suppression,Paired-pulse ratio, Presynaptic inhibition

In hippocampal slices, depolarization of a CA1 pyramidal neuron has been demonstrated to elicit transient suppression of inhibitory synaptic inputs to the depolarized cell (Pitler and Alger 1992). This phenomenon was referred to as "depolarization-induced suppression of inhibition" (DSI). There are several lines of evidence implicating an involvement of retrograde signaling in DSI (Alger and Pitler 1995). However, precise mechanisms of DSI, especially in regard to the retrograde messenger, are still unknown. Here we show that DSI is reproducible in cultured rat hippocampal neurons, suggesting that a dissociated cell culture system can provide a useful model system for the study on DSI.

Hippocampal neurons were prepared from newborn rats and cultured for 7-14 days. Monosynaptic inhibitory postsynaptic currents (IPSCs) evoked by a single presynaptic neuron were recorded from the postsynaptic neuron. For the induction of DSI, the postsynaptic neuron was depolarized to 0 mV for 3-5 s (Ohno-Shosaku et al. 1998). In about 40% of the inhibitory synapses, postsynaptic depolarization caused transient suppression of IPSCs (DSI) with a similar time course to that previously reported in hippocampal slices. The induction of DSI was Ca^{2+} dependent. DSI was associated with an increase in the paired-pulse ratio, indicating a decrease in the transmitter release. The inhibition of action potentials of the postsynaptic neuron by injecting a local anesthetic (QX314) intracellularly failed to prevent DSI. These results are consistent with previous studies with slice preparations.

We further examined the possibility that DSI can be induced at neighboring synapses on nondepolarized neurons as observed in Purkinje cells (Vincent and Marty 1993), which might be possible if a diffusible retrograde messenger is involved in DSI. For this purpose, we used a pair of neurons with an inhibitory connection from one cell to the other and an autaptic connection within the presynaptic neuron. IPSCs and inhibitory autaptic currents (IACs) were recorded from the postsynaptic and the presynaptic neurons, respectively. In some pairs, both IPSCs and IACs were transiently suppressed after postsynaptic depolarization (Fig. 1), suggesting propagation of DSI. These results are consistent with the hypothesis that a diffusible retrograde messenger is involved in DSI.

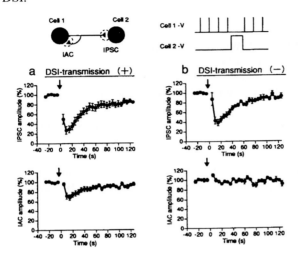

Fig. 1a,b. Propagation of DSI (depolarization-induced suppression of inhibition) in cultured rat hippocampal neurons. **a** In about half of the pairs exhibiting DSI, depolarization of a postsynaptic neuron (Cell 2) suppressed inhibitory autaptic currents (IAC) recorded from a presynaptic neuron (Cell 1) as well as IPSCs recorded from the depolarized neuron (Cell 2). **b** In some other pairs, the depolarization suppressed only IPSCs, but not IACs.

The present study demonstrated that DSI can be reproducible in a single pair of cultured rat hippocampal neurons. Using this culture system, we found that DSI can propagate from the depolarized cell to the other cell, suggesting an involvement of a diffusible retrograde messenger in DSI.

References

Alger BE, Pitler TA (1995) Retrograde signaling at GABA$_A$-receptor synapses in the mammalian CNS. Trends Neurosci 18:333-340

Ohno-Shosaku T, Sawada S, Yamamoto C (1998) Properties of depolarization-induced suppression of inhibitory transmission in cultured rat hippocampal neurons. Pflügers Arch 435:273-279

Pitler TA, Alger BE (1992) Postsynaptic spike firing reduces synaptic GABA$_A$ responses in hippocampal pyramidal cells. J Neurosci 12:4122-4132

Vincent P, Marty A (1993) Neighboring cerebellar Purkinje cells communicate via retrograde inhibition of common presynaptic interneurons. Neuron 11:885-893

DATE DUE
